Researching American Culture

A GUIDE FOR STUDENT ANTHROPOLOGISTS

Researching American Culture

Conrad Phillip Kottak,
Editor

Ann Arbor The University of Michigan Press

Published in the United States of America by
The University of Michigan Press and simultaneously
in Rexdale, Canada, by John Wiley & Sons Canada, Limited
Manufactured in the United States of America

Library of Congress Cataloging in Publication Data

Main entry under title:

Researching American culture.

Bibliography: p.
1. United States—Popular culture—Addresses, essays, lectures. 2. United States—Civilization—1970- —Addresses, essays, lectures. 3. Mass media—United States—Addresses, essays, lectures. 4. Ethnology—Methodology—Addresses, essays, lectures. 5. Ethnology—United States—Addresses, essays, lectures. I. Kottak, Conrad Phillip.
E169.12.R47 306'.0973 81-23175
ISBN 0-472-08024-5 AACR2

To G. Harvey Summ

Preface

I thank many people who, since 1977, have contributed to the organization and running of Anthropology 412 ("The Anthropology of Contemporary American Culture") and directly or indirectly to this book. Thanks first to Lebriz Tosuner-Fikes, whose idea it was to prepare a field research guidebook for undergraduates working on American culture. Tosuner-Fikes's experience as a 412 teaching assistant in the fall, 1978, term convinced her that college students needed to be reminded of terms and ideas that they had forgotten since introductory anthropology. She also thought that explicit instructions about all phases of planning and carrying out research would reduce the difficulties encountered by her students. Tosuner-Fikes drafted a grant proposal to the University of Michigan Center for Research on Learning and Teaching (CRLT), which, once awarded under my direction, has supported preparation of this volume. Thanks therefore are also due to CRLT. This CRLT Faculty Development grant permitted Tosuner-Fikes to prepare a guide, which is included as a chapter of this volume, during the summer of 1979. Thanks also to the Office of the Dean of the University of Michigan College of Literature, Science, and the Arts and the Department of Anthropology for stipends that helped cover preparation.

The graduate students who worked with me during the fall, 1979, term met with Lebriz Tosuner-Fikes and me to decide on a format. I gratefully acknowledge their advice and assistance, particularly in contacting the students whose papers are reproduced here. These 412 colleagues are Louise Berndt, Susan Gregg, William Meltzer, Linda Place, Lynne Robins, and Gale Thompson. I also thank 412 teaching assistants from past semesters who have made indirect contributions to the course and this volume. They include Jay Fikes, Philip Guddemi, Ellen Hoffman, William Kelleher, and Jeffrey Resnick.

A slightly shorter and different version of this book was originally published as a special issue (vol. 6, no. 1, 1980) of *Michigan Discussions in Anthropology*, the biannually published journal of the University of Michigan Department of Anthropology. All royalties for the current book have been assigned to the journal, to support future creative issues.

This book is designed for use in varied courses requiring research on aspects of contemporary American culture. These courses will include anthropology of contemporary American culture, introductory cultural anthropology, and field techniques. We anticipate that the book may also be useful in certain courses offered by sociology, American culture, and English departments.

I wish also to thank those who have helped me with the organization and preparation of this book, including Shafica Ahmed, Linda Krakker, Francine Markowitz, Mary Steedly, and the editorial staff of *Michigan Discussions in Anthropology*. Thanks also to the University of Michigan Press staff for its excellent cooperation.

Finally, this book is dedicated to my uncle, G. Harvey Summ, who, more than anyone else, encouraged my interest in using anthropological techniques and perspectives to study my own society. His enthusiasm helped me decide to offer the course out of which this volume developed. The essay reprinted here from my textbook, *Anthropology: The Exploration of Human Diversity* (2nd ed., Random House), was my initial foray into American culture, and it was written at Harvey Summ's house. Thanks, Harvey, for years of intellectual stimulation.

Conrad Phillip Kottak

Contents

Introduction

Conrad Phillip Kottak

This book is designed for use in undergraduate courses on the anthropological study of contemporary North American culture and in other courses, including introductory cultural anthropology, that require original field research on some aspect of the student's own society. The book grew out of a course offered each fall at the University of Michigan, Anthropology 412 (now 320), "The Anthropology of Contemporary American Culture." First offered during the fall, 1977 term, the course reflected my own developing interest in contemporary society, particularly in mass or "pop" culture. The course was also a response to the student enthusiasm that was evident whenever I used American examples to illustrate principles being taught in introductory anthropology. An original field research project was the main course assignment.

This book's aims are multiple. One objective is to show the value of anthropology in studying contemporary society. Another is to guide students in original research. Fifteen student papers have been included to illustrate actual student research accomplishments. This introduction and Part 1 (essays 1 and 2) review basic anthropological terms and procedures and offer detailed guidance for planning, carrying out, and writing up results of research projects. The introduction discusses anthropology's value for studying the contemporary United States and summarizes some organizing ideas, themes, and perspectives, while essays 1 and 2 guide the student from project planning through write-up of the completed research. The essays in Part 2 (3 through 10), mainly written by professionals, illustrate several anthropological methods useful in analyzing the contemporary mass media. Other research strategies, techniques, and subjects are examined in Parts 3 and 4, which contain most of the fifteen student research papers chosen for inclusion in this book.

A course that uses familiar cultural material to illustrate new techniques permits students to concentrate on learning methods and perspectives without also having to remember details of foreign ethnography. The contributions to this book by both students and professionals illustrate the course goal of teaching a variety of anthropological techniques, rather than any single approach. Among the research paradigms and analytic methods included are: emic and etic research strategies, structuralism, symbolic analysis, neo-Freudian analysis, componential analysis, sociolinguistic methods, content analysis, observation of public behavior, interviewing, and quantitative research design. This nondoctrinaire inclusion of multiple anthropological approaches is, I believe, one of this volume's unique features.

Another is its inclusion of a large number of original student research papers as the essays in Parts 3 and 4. These provide, quoting one of the anonymous prepublication reviews of this manuscript,

> excellent examples of how serious students can carry out interesting projects among friends, family, and co-workers, which yield useful insight into their own lives. They demonstrate the essential wisdom of paying close attention to narrowly defined problems which have larger contexts. They also, and perhaps most important, can demonstrate to students a standard of excellence for their own work.

After reading each paper several times, I am very impressed with college students' work and ability. Note that there were several student papers of similar quality that I could not include, either because of space limitations, or because the general subject matter was already represented. Very minor copy editing has been done on most of the student papers, mostly to correct spelling and punctuation, and to make reference style consistent. Unfortunately, because student contributors are now dispersed or have lost their list of references, we have been forced to include a few incomplete citations in the references. (The reference and bibliography style used in this volume follows conventions of the *American Anthropologist*, the main professional journal of our discipline.)

Another major course goal illustrated by the student essays is to help students think about their culture in new ways, to expose and call into question covert cultural biases and ethnocentric assumptions, to convert students from being simply natives into being native anthropologists. The job of the native anthropologist, discussed more fully in essay 3, is to use the detachment and cross-cultural perspective that training in anthropology provides in describing and analyzing familiar behavior. All anthropologists bring to whatever culture they study certain assumptions and biases acquired in their culture of origin. However, proper training increases introspection, objectivity, and cultural awareness, thus reducing the bias that one culture's ethnographer brings to the study of another. Native anthropologists attempt to combine an observer's impartiality with lifelong cultural expertise. As natives, they can draw on their knowledge of the native language, and their skills and understanding acquired during years of enculturation and formal study. As anthropologists, they can question native beliefs and categories and watch for regularities in behavior that natives are unaware of and might even deny if they were pointed out. Resulting explanations are often *etic* (phrased in observers' categories) rather than *emic* (acceptable to natives).

A more open, questioning, comparative, relativistic outlook permits the anthropologist, paraphrasing a tenet of Claude Lévi-Strauss's (1966,1967) structuralism, to delve beneath the conscious mind to discover the deeper structure of cultural behavior. Or, as Marvin Harris (1974) would have it, the scientifically trained student of culture can penetrate the "social dreamwork," the maze of sometimes contradictory, often inconsistent, frequently obscuring (and also, often enlightening) statements that natives make to describe, interpret, and explain their behavior. Thus, many of the contributors to this collection have rethought the American cultural category "trivia," viewing the phenomena so labeled as worthy of serious study, revealing pervasive cultural themes. Manifestations of "pop culture" that influence millions of lives, through conscious and unconscious enculturation, are certainly not trivial. Having been trained to be skeptical about (and thus to be attentive to, and to carefully examine) native categories, descriptions, and interpretations, anthropologists may examine the functions, meanings, and general cultural relevance of many cultural products, including those that natives view as "trivial" or as too obvious to require comment.

Selections in this book illustrate that many of the very same anthropological techniques that were developed in and for small-scale nonliterate societies can be applied to our own behavior today. American society is not just for sociologists, economists, and political scientists. Indeed, anthropology contributes a distinctive comparative and relativistic perspective. We have seen that anthropology helps students to become more objective about the supposed "naturalness" of their own customs, to become less ethnocentric, and to be more skeptical about certain beliefs and practices that are rarely questioned. Many students are initially extremely defensive about aspects of the American world view. I have found that the goal of objectivity is most difficult to reach when discussing certain pervasive values and beliefs. For example, students are reluctant to examine, to rethink, and to evaluate the common American belief that our living standard (diet, health, physical fitness, convenience, educational system, opportunity) is "the highest (or best) the world has ever known." And even after the energy crisis and years of ecological awareness, many students still hold onto the American core value (cf.

Arensberg and Niehoff 1975; Hsu 1975) that new resources can always be discovered or invented when current supplies are exhausted ("the frontier mentality," as articulated by the introduction to "Star Trek"—"space, the final frontier").

Among the earliest questions considered in a course on the anthropology of the United States are: Is there an American culture? What is it? When did it begin and how has it lasted? How has it changed? What are the key values and shared beliefs that we hold? Are they mutually consistent? Are certain beliefs we hold about ourselves better seen as cultural myths than as reflections of reality? One useful organizing theme is unity and diversity in American culture. In the essays in this volume, unity is explored in such common enculturators as public schools; national myths, symbols, and holidays; the nation-state; and, particularly, the mass media, which have had such a powerful and widespread impact on today's college students (and indeed on anyone who has grown up in the United States after 1940). This volume also covers expressions of social and cultural diversity, including those based on socioeconomic class, occupation, educational level, region, gender, age, ethnic group, race, family background, and urban, suburban, or rural residence.

Student Research Projects

This book attempts to prepare students for doing original research through specific guidance in project construction and execution (essays 1 and 2), illustration of many anthropological techniques, and inclusion of several actual student research papers. Prior to the field project, students in my course on American culture write a few short essays using anthropological methods to describe and analyze familiar phenomena. This is good practice for the larger project that follows. Three such essays (essays 5, 8, and 9) have been included.

The matter of what kinds of student research projects should be planned and approved poses a dilemma for anyone who wishes to instruct students in original research. If a large number of students is involved, it is hardly fair to unleash hordes of young and inexperienced scholars on a relatively small community to do ethnographic interviewing. At the University of Michigan we have tried to trouble our host community (Ann Arbor) as little as possible, for example, by discouraging students from studying the same places and groups. To safeguard the community and to promote good research, each student is required to submit a brief prospectus of the intended field project early in the semester. Teachers scrutinize these abstracts, discourage duplication of effort (with a file of previous years' projects), and discuss potential problems. Proper instruction in research procedures and in the ethics of social inquiry (reviewed by Tosuner-Fikes in essay 1) requires regular meetings between students and teaching staff. A student-teacher ratio of no more than twenty to one permits careful project supervision.

In deciding which projects to approve, instructors will be guided by enrollment in their class and the size of their community. In Ann Arbor, given large classes and a relatively small community, students are told to avoid projects involving interviews with strangers. Instead, they are advised to choose one of three kinds of research topics: (1) analysis of an aspect of the mass media, (2) study of behavior in a public place (without interviewing), or (3) study of a group to which the student already belongs. By encouraging students to focus on these areas, we are able to control the problem of students' bothering strangers. Thus, although interviewing, questionnaire construction, and sampling are discussed in essay 1, most of the student papers included in this volume emphasize the less intrusive types of research enumerated above. At small colleges, where classes have fewer students and interviewing is a possibility, teachers might want to assign this volume along with James Spradley's book *The Ethnographic Interview* (1979).

Readers will note, however, that some of the student projects included here did use interviewing of strangers. This kind of project was allowed if the proposal seemed particularly interesting, promising, and if the topic was not overly sensitive. For example, Suzanne Faber took advantage of her employment as a cocktail waitress at a local bar to investigate the relationship between income, tipping, and alcohol consumption. She asked 100 customers to indicate the level of their income on the back of their check, without telling them that she would try to correlate that information with the amount they drank and tipped. Although Faber used her workplace as a fieldwork site, her project did not strictly qualify as a study of a group to which the student already belonged. But her ingenious methods of collecting data, and the possibility of interesting conclusions, made her project well worth encouraging. Like Faber, Gail Magliano used her work experience, in an Ann Arbor financial institution, in her study of the expression of the American cultural value of right-handedness in local banks. Her project combined various techniques: observation of behavior in a public place, informal interviewing, perusal of company advertising, and study of a group to which she belonged.

Terrence O'Brien made use of a kinship tie in order to study a funeral home in suburban Detroit; he combined formal and informal interviewing with observation of behavior in a public place. He was able to correlate certain aspects of funerals with socioeconomic class and discovered regularities in Americans' reactions to death and funerals. Eric McClafferty's analysis of rituals of status elevation and reversal drew on his former membership in the high school swim team being analyzed, as well as on observation and talks with current team members.

Kenneth Schlesinger, in his research on racquetball players' reactions to lost points, observed public behavior, as did Mary Jo Larson when she studied talk by males and females in university classrooms.

Tina Van de Graaf and Francine Chinni used a tape recorder in their research on gender terminology among members of a local sorority. This student project was the only one based on formal interviewing of members of a group to which the researchers did not belong.

The student papers in Part 3, which are based mainly on observation of public behavior and study of groups with which the student was already affiliated, are preceded by three professional papers (essays 11 through 13) that make use of related techniques in similar microethnographic settings. Part 4 includes student research papers based on content analysis of the mass media. It begins with a professional paper, Maxine Margolis's demonstration of the far-reaching applications of a contemporary ideology, "Blaming the Victim," to females in the United States. The student papers by Rentz, Hesseltine, and Hill also demonstrate aspects of discrimination based on gender and race in our society. Patricia Rentz's quantitative analysis of the advertising and story content of eight "women's magazines" demonstrates that advertising and stories carry conflicting messages, the former telling women to remain in traditional roles, the latter emphasizing the contemporary woman's extradomestic life. Rentz's findings led to her identification of certain psychological conflicts that women experience because of this contradiction in the mass media's tacit enculturation. In a related study, Patricia Hesseltine watched every minute of commercial television for a twenty-hour period and charted the occupations of television women, as well as their authority and effectiveness (including commercials about superwomen who can "bring home the bacon, fry it up in a pan, and never, never, never let you forget you're a man"). Today's feminist consciousness leads us to believe, perhaps, that TV is showing fewer home-oriented women, and more women working for cash. But Hesseltine's detailed analysis says no: 83 percent of the women were homebodies.

Focusing on four daytime soap operas, Christina Hill finds unrealistic portrayal of blacks; she compares actual black residential patterns, employment histories, and occupations with those of the soaps' black characters and concludes that daytime viewers are being deluded into the belief that blacks can easily succeed in today's America, given achievement motivation. Hill contends that this belief diverts viewers' attention from social problems and reaffirms Americans' traditional and erroneous ideas that the poor are poor because they are lazy and that anyone who wants to work can find a job—beliefs that express core values of individualism and individual achievement.

The final student essay, by Fermin Diez, stands alone. It is a foreigner's reaction to the popularity of sports, particularly football and baseball, in the contemporary United States. Diez's research drew on his own participant-observation as an outsider confronting American culture, on the anthropology and sports literature, on the presentation of sports in the mass media (particularly television), and on informal conversations with Americans. Diez's essay relates the American preoccupation with sports to the American value system, as both are perceived by an outsider.

Thus, the essays offer glimpses of many expressions of unity and diversity in American society. This book is intended as a guide to anthropological research methods, interpretations, and explanations and as a sampler of what serious undergraduates can accomplish when they do research in their own society. It certainly does not purport to be a complete and comprehensive anthropological treatment of contemporary American culture. It offers encouragement and illustration rather than the final word.

Researching Contemporary American Culture

The essays in Part 1 have two objectives: (1) to justify the value of anthropological perspectives and techniques in studying American culture, and (2) to offer detailed guidance for student research projects. Essays 1 and 2 review anthropological terms relevant to research and make specific suggestions about project planning, research, and write-up. Part 1, therefore, serves as an introduction to the specific studies of mass media, public behavior, social principles, and cultural beliefs that follow. The first half of this book (most of the essays in Parts 1 and 2) has been written by teachers for students, while the second half consists mainly of research papers written by undergraduate students.

Tosuner-Fikes's guide, which students should consult before and during project planning and actual fieldwork, reviews key anthropological concepts and techniques, along with ethical considerations that all anthropologists should bear in mind during their research. This guide discusses research from start (topic selection and proposal) to finish (write-up as a paper). The student essays later in this volume exemplify successful final products.

1 / A Guide for Anthropological Fieldwork on Contemporary American Culture

Lebriz Tosuner-Fikes

This guide is designed to aid undergraduates doing original fieldwork in anthropology. The course assignment to design and conduct an anthropological research project provides a unique and valuable learning experience. Students should benefit from their work in various ways: acquiring new information about the meaning of culture and society in relation to themselves; learning how to do fieldwork and to evaluate research results; and producing original work that enhances their academic abilities as well as their creativity. The research experience is a learning process with immediate and pragmatic consequences. This guide is intended to assist you in achieving these goals.

Anthropological Concepts Used in Research

Anthropology sets itself apart from other social sciences by focusing on the study of culture. The term *culture* has many meanings, and it is best to choose a broad definition, such as Kottak's (1978*a*:536):

> That which is transmitted through learning—behavior patterns and modes of thought acquired by humans as members of society. Technology, language, patterns of group organization, and ideology are aspects of culture.

Culture can be studied at two conceptual levels—the universal and the particular. For example, the family unit is an organization of related individuals. This unit is found all over the world; "the family" is universal. But the special and specific rules that organize the unit are not universal. For example, in some cultures, but not others, a cousin can be seen also as a sibling. Therefore, although "the family" is universal, its composition is particular. We understand similarities by understanding the differences. By accumulating field studies based on comparison and contrast, we can better understand the commonality of the meaning in "family." Thus, the two conceptual levels of culture are interrelated.

Aspects of Culture

We share mental categories with members of our own culture. We do this because of our common *ideology*. Ideology includes "values, norms, knowledge, themes, philosophies and religious beliefs, sentiments, ethical principles, world-view, ethos, and the like" (Kaplan and Manners 1972:112). By means of an ideological system we build an identity and make ourselves secure through knowing what our proper code of behavior is when relating to others. Consequently, we create a code which defines a reality shared by its participants. Thus we give meaning to our environment; we maintain an existence that makes sense. Of course, what that "sense" means is culturally defined.

Each culture's ideology requires special appreciation by the analyst. The observing anthropologist must be self-critical before beginning research. Anthropologists must admit and examine their own biases or else they cannot trust their data and findings, since their cultural preconceptions might keep them from constructing a complete and accurate account. Analysts must constantly remind themselves of the importance of being objective, and make a conscious and persistent effort to adopt a relativist's position. Cultural relativism is "the attitude that a society's customs [must] be viewed in the context of that society's culture and environment" (Ember and Ember 1973:384). It is vital that the analyst assume this attitude when investigating *any* kind of cultural behavior. By being relativistic we can better understand the people we study as they perceive their own existence (Bronislaw Malinowski strongly emphasized the importance of grasping the native's point of view in his introduction to *Argonauts of the Western Pacific* [1961]). We can then assume a role that interprets the material within a larger context.

The difference is one of an *emic* versus an *etic* perspective. These are complementary to one another; neither can be fully applied or appreciated on its own. The emic/etic research strategies are summarized by Kottak (1978*a*:5–6) as follows:

> To study different cultures, anthropologists have advocated two approaches, the emic (actor-oriented) and etic (observer-oriented). The emic approach views a culture as mental or ideational and

> assumes that it can be described only by getting into the heads of the people studied. . . . The anthropologist seeks the "native viewpoint" and relies on the culture-bearers, or actors, to judge whether something they do, say, or think is significant or not.
>
> The etic . . . approach implies different goals for the anthropologist. When describing, interpreting and analyzing a culture, the anthropologist relies on his or her own extended observations and gives more weight to the trained scientist's criteria of significance than to those of the culture-bearers. Choice of the etic research strategy rests on the assumption that, as a trained and objective scientist, he or she can take a less involved, more impartial, and larger view of what is going on.

A nonethnocentric relativist position must also be applied while doing intracultural studies (e.g., of the contemporary United States). Even if the analyst is a native of the culture under examination, he or she must continue to work with an emic/etic perspective during the analysis of cultural parts.

Anthropologists who are interested in intracultural variation look at the broad range of possibilities in *enculturation*, the process whereby individuals learn a particular cultural tradition. We must observe how people relate to one another within the boundaries set by their ideology. What kinds of behavior maintain the culture and what kinds enable change to occur? We must learn to identify contradictions between belief systems codifying behavior, and the behavior that we actually observe. If there is a consistency between ideal and observed behavior then we must strive to understand and explain the conditions permitting such consistency. Thus, anthropology is interested in how some behavioral sequences produce new behavior whereas others perpetuate the same behavior, and how behavior is passed on from generation to generation (cf. Bateson 1958, 1972). Anthropologists are interested in observing, recording, and understanding value systems that define a culture and consequently affect the formation of personalities. Anthropologists discern a relationship between individual and group behavior and ideology; what they attempt to reveal are the mechanisms by which such an interaction is made possible. In order to do so, we need to look at other relevant cultural features.

According to Ember and Ember (1973:58), *ethnohistoric research* is composed of "studies based on descriptive materials about a single society at more than one point in time," and providing

> the essential data for historical studies of all types. . . . Ethnohistorical data may consist of sources other than ethnographic reports by anthropologists; the materials of ethnohistory may include accounts by explorers, missionaries, traders, and government officials.

Ethnohistory, therefore, refers to the history of a particular culture or ethnic group. Consideration of historical trends, both regional and global, is important. A specific change within a culture must be seen in relation to broader aspects of change. By studying ethnohistory one learns why certain traits, beliefs, and artifacts are retained, why others are not, and why still others are adapted to new cultural forms. This provides a total rather than a partial understanding of culture and behavior.

Along with history, *cultural ecology* plays a very important role in shaping many behavioral patterns and affecting types of cultural styles. Cultural ecology refers to interrelationships between a human population's physical environment, natural resources, and cultural traditions. The concept of cultural ecology was first introduced into anthropology by Julian Steward (1955), who related many aspects of cultural variation to the adaptation of societies to their environments. He believed that physical as well as social environments affect the development of culture traits. Thus,

> individuals or populations behaving in certain different ways have different degrees of success in survival and reproduction, and, consequently, in the transmission of their ways of behaving from generation to generation. [Ember and Ember 1973:50]

This view tries to establish a relationship between the behavior of people in a particular culture and the resources found in their environment. For example, cultural ecology aims at explaining social organization, task assignment by gender, and cultural values governing relationships among group members as well as between groups, on the basis of environmental variables.

In addition to ideology, mechanisms of enculturation, ethnohistory, and cultural ecology, another indispensable part of any study is the analysis of *social structures* and their *functions*. But what are social structures? Are they mental concepts, as some anthropologists (e.g., Lévi-Strauss) believe they are? Or are they universal institutions marked by terminology (e.g., Radcliffe-Brown 1952)?

The point that is common to all definitions of social structures is that they provide us with the means and arena where "learning" takes place, and they are a very real part of the complex society in which we live, and on which this research will be conducted. Our society does make definite distinctions in separating institutions from one another—e.g., the institutions of the family, medical system, government, educational system—and it makes these separations by assigning different functions to each. While observing these institutions, and understanding how they teach cultural ideals, it is interesting to note that as we learn values we simultaneously carry out the values we are aiming to adopt.

Furthermore, a common learning style permeates all the institutions that teach different sides of cultural norms. For example, in the family, parents teach; in educational systems, teachers teach; in the military, officers teach. In each institution learning passes "down" from a "giver role" to a "receiver role." In our cultural perspective, learning occurs by means of information going one way between two roles. The manner in which those roles are positioned reflects the hierarchy of cultural values (authoritative/subservient); the direction of information flow is down.

We are continually confronted by this model in every institution where learning takes place. Learning models are cultural products determined by multiple factors (two of these being history and cultural ecology), affecting the design of social, economic, and political systems. Such systems are held together by and represent an ideology. Each ideology reflects cultural norms that appear as repetitive themes in all aspects of culture. For example, in the capitalist system, competition is a repeated theme of behavior. This sort of theme has been termed a "core value" by Arensberg and Niehoff (1975).

Our responsibility, as anthropologists at work, is to penetrate the mystique of superficial variables that camouflage the deep structural variables that determine and produce behavior. In this manner we can appreciate the meaning of culture, as well as describing accurately a particular culture and analyzing its composition.

Analytic and Relational Modes

In solving a research problem, there are alternative ways of analyzing behavior. Although it is reasonable and analytically convenient to emphasize one approach over another in a given analysis, this is not to say that there is only one correct way to view a problem. It is possible to manipulate the same problem so that different anthropological tactics can be used, each focusing on a different point of interest. It is also possible to combine various approaches, thereby achieving a multifaceted problem-solving attitude.

In pursuing research, anthropologists recognize two modes of cognition: the analytical and the relational. By taking an analytical cognitive style, the analyst brings the specific topic of interest under microscopic examination. Each aspect of the topic is scrutinized. The problem is clearly understood in terms of its origin, its persistence, and its future direction. The way in which interaction occurs between members of the group under study is clearly defined, and details are presented in the process of describing behavior as well as interpreting it.

For example, to be analytical about competition in education, one might first study the classroom. Here the analyst could describe the spatial (symbolic) design of the room and how it contributes to the types of interaction that occur during the class. In relation to space and design, what do the artifacts project as a message? Evaluation of the time reflected in the class period is necessary. Often the time factor creates tension about performance. The ability to deal with stress in competition and success in competing are closely linked to the ability to be productive under time constraints. The kinds of talking between students, and between them and the teacher, must be accounted for. Body language is another important part of classroom communication. Finally, the contribution of all these points to the actual learning experience must be understood. In covering these areas, we are able to get a clear picture of how competition operates in the classroom (see, e.g., Henry 1975; Fikes 1978).

Once competition in the classroom has been sufficiently analyzed (so that the reader is convinced of the researcher's point), the role of competition beyond the classroom must be examined. This involves the relational style. The educational system itself must be understood in order to relate classroom patterns to the rest of the institution. And in order to fully understand how the educational system produces competition, it must be viewed within its total cultural context. In examining the acquisition of knowledge, the analysis moves from the particular to the general by relating behavior in a part of culture to the totality of the culture itself. A behavioral theme that ultimately reflects a predominant or core value in a cultural system is traced through levels of an institution. Classroom competition is understood in relation to other aspects of the educational system, which in turn can be appreciated fully only in relation to other aspects of culture. Behavioral themes are discovered at the analytical cognitive level, but the completeness of their meaning can be revealed only through the relational cognitive style.

Ethnography and the Research Assignment

Having reviewed some basic anthropological concepts, we may turn to the role of the researcher and the purpose of the research assignment. Essentially, the research assignment expects the student to produce a "mini-ethnography." What does *ethnography* mean and what is its purpose? According to James Spradley (1979:3), ethnography is simply "the work of describing a culture." He continues,

> The essential core of this activity aims to understand another way of life from the native point of view. . . . Fieldwork . . . involves the disciplined study of what the world is like to people who have learned to see, hear, speak, think, and act in ways that are different. Rather than *studying people*, ethnography means *learning from people*.

The student's ethnographic report will be a detailed account, based on careful observation in a particular cultural setting. It will take into account both native (emic) opinions and the student's own (etic) understanding of the behavior being studied. Certain ethnographers believe that the purpose of an ethnography is simply to describe, analyze, and document a culture. Others believe that the ethnographer's responsibilities go beyond these steps, and that the data should be used to take action in order to solve problems. Such basic assumptions of the ethnographer must be considered when reviewing data; they will undoubtedly affect the product. Because it becomes very difficult to separate the two roles when doing ethnography in complex societies, most ethnographies of American culture seem to carry both a social and a moral message, and it is becoming increasingly difficult for anthropologists to divorce themselves from applied work. According to Kottak (1978*a*:496), there are

> fundamental differences between the attitude of a majority of contemporary anthropologists and the viewpoint that holds that anthropologists' ethical and value judgments should be totally distinct from their scientific work.

And according to Spradley (1979:16), ethnography is for understanding the human species, but also for serving the needs of humanity. One of the challenges facing every ethnographer is to synchronize these two aims of research.

Ethics

The student researcher, as a fledgling anthropologist at work, should also be aware of the code of ethics observed in anthropology. Before designing a project and conducting fieldwork, you should be aware of your responsibilities as an ethnographer in relation to project, subjects, self, and the discipline. A good beginning is a review of the first four items in the code of ethics of the American Anthropological Association, drafted in 1970 under "AAA: Principles of Professional Responsibility," as paraphrased by Kottak (1978*a*:496).

1. *Responsibility to Those Studied*
   Anthropologists' paramount responsibility in research is to those whom they study. Anthropologists must do everything they can to protect informants' physical, psychological, and social welfare, and to honor their dignity and privacy. If interests conflict, these people come first. The rights, interests, and sensitivities of those studied must be protected. Specifically, anthropologists should make known to informants the aims and the anticipated consequences of their investigation; they should ensure that informants preserve their anonymity in all forms of data collection. Individual informants should not be exploited for personal gain. Anthropologists must anticipate and take steps to avoid potentially damaging effects of the publication of their research. In accordance with the AAA's official disapproval of clandestine or secret research, no reports should be provided to sponsors that are not also available to the general public.
2. *Responsibility to the Public*
   Anthropologists owe a commitment to candor and to truth in disseminating their research results and in stating their opinions as students of human life. Anthropologists should make no secret communications, nor should they knowingly falsify their findings. As people who devote their professional lives to understanding human diversity, anthropologists bear a positive responsibility to speak out publicly, both individually and collectively, on what they know and what they believe as a result of their professional expertise. They bear a professional responsibility to contribute to an "adequate definition of reality" upon which public opinion and public policy can be based. In public discourse anthropologists should be honest about their qualifications and aware of the limitations of their discipline's experience.

3. *Responsibility to the Discipline*
   Anthropologists bear responsibility for the good reputation of their discipline and of its practitioners. They should undertake no secret research or any research the results of which cannot be freely derived and publicly reported. They should avoid even the appearance of engaging in clandestine research by totally and freely disclosing the objectives and sponsorship of all research. They should attempt to maintain a level of rapport and integrity in the field such that their behavior will not jeopardize future research there.
4. *Responsibility to Students*
   Anthropologists should be fair, candid, nonexploitative, and committed to the welfare and academic progress of their students. They should make them aware of the ethical problems of research.

Although student researchers are not officially anthropologists or teachers, their original fieldwork will nevertheless have some teaching implications. Regardless of scale, the research results will provide additional information on a particular subject and will be passed on to others in some form. Perhaps some students will produce results that can be published (such as the papers in this volume) or applied to policy making. In any case, information is transmitted, and researchers must assume responsibility for their work as a teaching tool.

The Research Project Design

Project design (the research proposal) is the most critical part of research. The success of the research, in terms of its progress and completion according to schedule, the degree to which the researcher has control over the direction the work will take, and the kind of information that will be provided all depend primarily on the initial research design. The design is an outline of your ideas. It projects your goals, reflects knowledge you have, and influences the kinds of data that you will acquire during fieldwork. It announces, in a nutshell, your reasons for doing your particular work, what you expect to find in doing it, how you will conduct your work (in detail), and why it is relevant to anthropology. Before writing your research proposal, be sure you have thoroughly examined each of these points.

The project proposal is a necessary part of "preaction research." It presents your design in written form. It is an abstract of what to expect from your project. Therefore, your proposal must be carefully planned and thoughtfully written. It should convince readers of the possibilities and relevance of your research. The best way to go about writing up a proposal is by switching to another perspective, the reader's (normally the instructor's). By assuming another person's perspective and critically reviewing the proposal, you can predict what the reader will be looking for. Crucial questions for the reader include:

1. Is the research problem or topic sufficiently important to be encouraged?
2. Is the hypothesis that the student intends to test a reasonable one? Is the student's viewpoint plausible?
3. Is the research topic too sensitive or controversial for an inexperienced researcher?
4. Will the research intrude on the privacy of others? If so, is this intrusion warranted?
5. Does the student have the skills and background information necessary to do the research, or can they be acquired in the time available?
6. Has the research topic been narrowed sufficiently so that the student can carry out the investigation in the time available?
7. Are the methods outlined by the student appropriate for the proposed topic?

Of course, you as a researcher must first satisfactorily answer these questions. The initial step in accomplishing this goal is to focus on topic selection.

Choosing a Topic

When selecting a topic consider two points: choose something that interests you and contributes to your knowledge; and choose something with easy accessibility. As far as your interest is concerned, pick a topic that you are already familiar with and want to expand upon, or pick a topic that you know little about but that might prove useful to you, so that you can use this opportunity to learn about it. As far as accessibility is concerned, pick a topic that you can conveniently investigate on a regular basis, with a consistent sample.

For example, if you select a project that requires that you be outdoors most of the time, make sure you are the outdoor type. Remember that during most academic terms seasons change. Don't let the weather create problems for you. Another consideration is distance. Make sure your investigation site is not out of your way. Or, if it is, consider difficulties with transportation. Remember that projects of this sort require you to collect data on a regular basis within boundaries that you have defined. It is inappropriate to collect data that do not represent the same sample that was originally indicated in your proposal. It is also inappropriate to collect a chunk of data in one shot and then use it for a project that is supposed to cover a more extended period of time.

After having considered the practical implications of these points, one student in the fall, 1978, class decided to do his research on behavior among college students at a particular discotheque. He knew that disco dancing was a popular activity but, not being a dancer himself, he had no clear idea about the meaning of the disco craze of 1978–79. He simply took advantage of the opportunity to learn about the disco scene and its relevance to American culture. Also, since he was not at all the outdoor type, he needed an indoor project. He needed no transportation to the disco since it was within walking distance. Furthermore, he attended school and worked, so that his only available time for fieldwork (as well as to "get out") was on a Saturday night. For a designated period, he regularly went to the same establishment. As it turned out, he collected high-quality data without resenting the demands of disciplined fieldwork. During this time, he also went through a critical analysis of his own behavior in relation to others within the context of a college experience. He reevaluated the function of the educational process in terms of the behavior patterns invoked by it. He saw the development of these patterns within the context of the larger social system and linked them to American cultural values. In view of all this, his research experience was a positive one, and he gained an increased appreciation for anthropology.

Of course, the way in which this student's project took shape was predictable from the outset in his proposal, which was well planned. The importance of having clear goals and concrete research methods cannot be overemphasized.

Sample Proposal

Let us analyze in depth a proposal presented by a former student (from fall, 1978). This student was a senior concentrating in sociology, who was preparing to enter medical school. The design of his project fulfilled multiple needs. He learned how to use the anthropological method; he acquired additional knowledge about culturally defined behavior in the medical field; he added the newly obtained information to his previous research and was able to adopt a cross-cultural perspective (he had already done similar research in England and Sweden). Most important, he was able to combine a humanistic and a scientific understanding of patients' problems, thereby enhancing his future career effectiveness. The student's unedited proposal illustrates how these ends were achieved:

(1.) In America today there seems to be a general public dissatisfaction concerning the practice of medicine. (2.) Many times this displeasure stems from the doctor-patient relationship and what preconceived expectations the patient may have before entering the relationship. (3.) How the doctor responds to the patient then becomes an important factor. (4.) Adequacy of the explanation of illness and warmth generated in the interaction are major determinants that have been found to influence the patient's satisfaction. (5.) Although these were the major determinants of satisfaction illustrated by Korsch and Negrete in a study done at the Children's Hospital of Los Angeles in 1968, other studies have resulted in different conclusions as to what are the most important factors leading to patient satisfaction.

(6.) Judging from the conflicting opinions in these studies as to what aspects lie behind this medical problem, I feel that further research is necessary. (7.) I believe that the social class of the patient as well as whether he/she is an urban or rural resident will influence which aspects will be of greatest importance to the patients in evaluating the doctor. (8.) These different personal characteristics will shape patients' expectations of how a doctor should act, and patient satisfaction will be contingent upon the extent to which these perceptions are met.

(9.) I plan to study this theory by talking with people door-to-door as well as on the street. (10.) I propose to use the questions listed below and select people from different areas of Ann Arbor and its vicinity. [There are certain problems with such an interviewing strategy. See the introduction.—Editor] (11.) Of course, to carry out my plans I must keep in mind the relevance of my findings. (12.) I realize that my field work could only at best be considered an initial probe, but I believe that it should allow me to gain a greater understanding of the problem, its implications and possible solutions that a future, more structural and tested study may reinforce. (13.) I may also be able to draw cross-cultural comparisons by utilizing the data I collected this summer in England and Sweden.

Proposed Questions
Occupation:
Age:

1. Do you have one personal doctor?
   Yes ______ /No ______
   Are you happy having one/many doctors?
   Yes ______ /No ______
2. Are you satisfied with your doctor(s)?
   Yes ______ /No ______
3. What are the qualities about your doctor that you appreciate?
4. Are there any other qualities that you feel a doctor should have, that your doctor does not have?
5. What characteristics in your doctor do you dislike?
6. I am new in this area and looking for a good doctor. How would you describe your doctor to me?
7. Do you feel that your doctor gives you an adequate explanation of your illness and his proposed treatment?
   Always ______ /Sometimes ______ /Never ______
8. What level quality of care do you believe you are getting?
   excellent ______ /fair ______ /don't know ______ /
   good ______ /poor ______
9. Do you ever have an idea of what is wrong with you before you go to the doctor? Yes ______ /No ______
10. Does your doctor have an appointment system?
    Yes ______ /No ______
    How long do you usually have to wait to see your doctor?
    On the average, how long is your visit with the doctor?
    Do you ever want more time and are denied this?
    Yes ______ /No ______
11. What made you choose this particular doctor?
12. Would you go to a doctor for an emotional, nonphysical problem? Yes ______ /No ______
13. Are you satisfied in general with the health care system?
    Yes ______ /No ______
14. If you have any additional comments, please use back of question sheet.

After having read the first paragraph, two points are clear to you as the reader: (1) I have understood the purpose and informative value of this study; and (2) I want to go on reading. In deciding to read on, you might have the following thoughts: I am of this culture and I do rely on medical doctors for health care. Let me remember how my relationship is with them; let me remember if I have heard many of my friends complain. Well, yes. Most people seem to be discontented. I wonder why? This person says that the discontent may be due to the way doctors relate to patients and to the failure of treatment to fulfill the patient's expectations. Of course, there are different reasons for a patient's disappointment. This person had appropriately narrowed his research topic by mentioning two variables (social class and rural/urban residence) that might produce differences. I wonder if there is any truth in what he says?

In order to understand what features of the proposal contributed to your decision to continue reading, let us review the first paragraph. Sentences 1 and 2 orient you. They may clarify the problem to be studied and suggest a culturally defined group of people (the general public) to whom the study would be relevant. Sentences 3 and 4 focus upon the variables that create the problem. Sentence 5 provides a reference for the support of the opinion presented in the two previous sentences.

Using a reference at this point is an important and appropriate tactic—a good writing technique. In this example, it shows immediate support for a general statement that presents the problem to be investigated. An appropriate reference helps convince your reader of the plausibility and significance of a problem. In general, while writing the proposal and your final project report, be sure to include references when appropriate. That is, use them if the statement presented is not originally yours, or if the statement expresses your opinion, which also agrees with another's who has tested (or researched) the problem and thereby documented your statement's accuracy. If reading assignments are given in conjunction with the research project, it is advisable to insert them as references where they apply. This way your reader is more apt to believe that you know what you are talking about. This also demonstrates that you have understood and applied the readings on the subject, thus supplementing your own findings. In general the more support you present, the stronger you make your case.

However, do not insert references haphazardly just to make an impression. An inappropriate reference will bring doubt into your reader's mind; he or she will question the sincerity of your work, or perhaps your logical ability. Often, such an attitude will cause your reader not to take your work seriously. Simply include those readings that agree with your perspective, or even those that do not, if you refute them explicitly. In this way, you show your reader that you are familiar with the topic, so that he or she will be convinced about your presentation. Remember, you are conveying a message to your audience that you have reviewed the literature (pro/con), and that you know what you are out to prove. Show your reader you have confidence in yourself and your work, so that the reader will, too.

To return to the sample proposal, the second half of sentence 5 indicates that there is a conflict of opinion on the reasons causing the social problem. In view of this, in sentences 6 and 7 the writer states that he believes further research must be done on the matter. He suggests alternative social variables that he think may affect the outcome. In sentence 8, with the background information thus far presented in mind, he presents his version of the underlying causes of the problem. In other words, here we have the researcher's hypothesis.

Data Collection: Hypothesis Framing

So far, the researcher has wasted no time in getting to the point of his project. In the next part of the proposal he presents his approach to research data. Before going on further with the proposal analysis it is important to review briefly the reasoning involved in determining statements of purpose. In sentence 9, the author refers to his proposition as "theory"; however, I would regard it as a "hypothesis." These words have similar meaning and are often used interchangeably. Their usage is somewhat ambiguous and may depend on the user, but defined more precisely, a hypothesis is a statement based on previous information (be it tested or merely assumed). It proposes a course of action indicating a testing process that will prove or disprove the claims made by the researcher. A theory, on the other hand, is the statement that is made subsequently. While one is made prior to testing the problem (hypothesis), the other is made after testing it (theory). As Pelto (1970:16) similarly states:

> Successful testing of the hypothesis adds to a body of theory or elaboration of a model; which in turn leads to re-examination of "the real world" and the techniques of observation.

Concerning hypotheses, Pelto (1970:206) provides an outline of the procedures used in testing the researcher's proposition.

1. Statement of a problem involving two or more variables and their interrelationships.
2. Presentation of research hypotheses.
3. Statement of methods of research.
4. Statistical tests of significance concerning the hypotheses.
5. Acceptance (on the basis of statistical tests) of the hypotheses.
6. Explanation concerning one or two relationships that "didn't turn out" in the predicted manner.
7. Statement concerning the theoretical advances achieved.

Of these, points 1 and 2 have already been included in the student researcher's proposal. Point 3 is presented in his third paragraph (to which we will return shortly). Points 4 and 5 are optional, since quantification is not necessary in all ethnographic fieldwork. Point 6 is very important when one is evaluating the project in relation to the initial proposal. Unanticipated findings sometimes prove most significant. The researcher must then understand and clarify why certain aspect(s) were overlooked during research design and data collection.

Kaplan and Manners (1972), in their elaboration of "theory," best illustrate the relevance of point 7 to anthropology. They remark that "in general we *construct* or devise theories," and they point out the importance of theory building.

> Theory-building and framing of explanations have important pragmatic implications. Being able to predict correctly allows us to anticipate events and thus to prepare for them. But if we know why we are able to predict correctly, we are provided with a mechanism by which we may also be able to intervene in events and exert some control over them. [P. 18]

The information covered here concerning "theory" is, I think, sufficient to make a review of the proposal in question possible. It should also help you understand the meaning of the "theoretical part" of the project. Now, let us return to the sample proposal. The section to be evaluated deals with research methodology.

Sampling

As can be concluded from sentences 9 and 10, the researcher will be using a questionnaire to collect data, and the manner in which collection will occur is by means of random sampling. Random sampling can be used in anthropological research, but there are problems involved because it is easy to confuse it with "haphazard sampling" (cf. Pelto 1970:163–164).

The sample is a smaller study group drawn from the larger population about which the researcher wishes information. In random sampling, each member of the larger population has an equal chance of being chosen as part of the sample. For example, the task might be to choose a sample of 25 from a department store staff numbering 125. A random sampling procedure would ensure that all 125 employees had an equal chance of being chosen, and that no systematic bias favored any subgroup, for example first-floor salespeople. One way to choose a random sample is to assign numbers to each member of the population, write each one on a piece of paper, and draw them from a bowl while you are blindfolded. Another is to consult lists of random numbers, found in virtually any statistics textbook. The absence of systematic bias is the key to any random sample, and the laws of probability, on which most tests of statistical significance are based, can only apply to random samples. [Tests of statistical significance were not required of the students whose reports are reprinted in this book.—Editor] A daytime study that questioned fifty people on a city street about their opinions of the American medical system would be haphazard rather than random. Its results could not be generalized to "the American public" since many groups would be systematically excluded—for example, shut-ins, rural residents, people at work. There can be no assurance that haphazard samples accurately sample larger universes.

Ethnographic Techniques

In collecting research data anthropologists traditionally use two techniques: participant-observation and ethnographic interviewing. In the former, the researcher both participates in and observes the events and behavior being studied. In ethnographic interviewing, he or she asks specific questions to people (informants or respondents). In the research proposal under consideration, participant observation cannot be used, since the student cannot observe several cases of physician-patient interaction, and since there are limits to the number of doctors' offices he can visit as a patient. The reasonable alternative is to use the ethnographic interviewing technique.

Interviewing. The design of the interview is geared toward acquiring information that provides ethnographic data. Interview design must include considerations of location, time spent in interviewing, kind of questions asked, and the types of informants to be interviewed. For a detailed discussion of these points see J. P. Spradley's book *The Ethnographic Interview* (1979).

Spradley (1979:25) states that

> ethnographers work together with informants to produce a cultural description. . . . The success of doing ethnography depends, to a great extent, on understanding the nature of this relationship.

He also mentions the role of the informant, and the various contributions an informant makes to the ethnography. Of these, I think the following two are the most valuable: "Informants provide a model for the ethnographer to imitate" so that the ethnographer can learn to adopt the emic perspective, and "informants are a source of information."

For these reasons, when selecting an informant, make sure you pick an individual who is thoroughly enculturated in your topic of investigation. (Note that this method of choosing an informant will not be used in projects that require formal sampling. Here I assume that you are working like the traditional ethnographer in a small community, who chooses particular informants who are cooperative and knowledgeable about certain topics.) It helps if the individual is talkative and eager to cooperate. However, silence is its own message; so if you are finding it difficult to get individuals to talk about an issue, try to figure out why. Frequently the best informants are those who are themselves participants in the subject matter you are investigating; often they love to talk about their past experiences. However, bear in mind that retrospective information may be very different from that which is gathered during the experience itself, and you will not be able to check the accuracy of your informant's statements against your own observations. Make sure your involvement in the subject of investigation, as well as that of your informant, is clear to *you*. You must define your research so that you do not lose sight of the etic perspective.

Also, it is of crucial importance to choose individuals who have the time to be interviewed. You should know beforehand whether your schedule coincides with your informants'. Cooperative timing and following through with commitments are important considerations in selecting your informants.

Be alert to the kind of information your informant is conveying to you. Is it his or her analysis of the topic or a pure description of it? Is this the informant's perspective or that of someone else? Is your informant being honest, or manipulating the material to impress you (are there ulterior motives involved)? Any time there is a sense of discomfort in the atmosphere, try to pinpoint its cause. Sometimes minor details can lead to major irritations. Whenever possible try to eliminate "bad vibes." However, sometimes the interviewer and interviewee just "don't hit it off." If you must, let go of an informant, and look for one who is more compatible with you.

When you have located the appropriate informant, one who is cooperative, knowledgeable, and reliable, there are various considerations for the interview itself. The place where the interview is to occur must be selected carefully. It should be convenient both to researcher and to informant. It should not be too noisy, because the ethnographer must pay close attention to the interview, and because of requirements of tape recording, if such is to be used.

If your informant refuses to allow recording, you are obliged to comply with his or her wishes. If recording does occur, unless the informant has granted permission to be quoted, make sure the material appears in such a way as to ensure privacy. You must maintain your informant's confidence at all times by remaining loyal to his or her best interests. It is essential that you explain fully and clearly the purpose of the interview. Of course, if the informant wishes, he or she may have access to the recorded interview at any time. It is important to make sure your informant feels that he or she can trust you. If private information is to be discussed, make sure the chosen place for the interviewing does not expose the informant. It is important to avoid making the informant feel self-conscious and embarrassed. Show an interest in the session and let your informant know you are curious, but never be openly judgmental. Remember, your role is one of documenter.

Querying informants: questionnaire construction. When conversing with your informant, you will have to "play it by ear" in determining the appropriate conversational style. Some informants are very comfortable with the formal questionnaire, such as the one accompanying our proposal. It does not intimidate them or belittle their competence; they know exactly what they are supposed to answer. These informants do not like general (open-ended) questions because they may fear that they are not answering with what the researcher is looking for. On the other hand, some informants feel restricted by formal questionnaires. Formal questionnaires sometimes do not allow for the informant to explain his or her perception of the problem adequately. For these individuals a broad question, such as: "There seems to be a problem in the kind of relationship between patient and doctor during a medical examination. How do you feel about that?" might provide a better format for the informant and consequently higher quality data for you.

Frequently researchers become impatient because informants answering broad, open-ended questions do not seem to be directly responding to the questions asked. If this occurs, and you have listened enough to realize you are truly wasting your time (in spite of the different interrogative tactics you used), then try to bow out as graciously as possible without offending the interviewee. However, impatience sometimes can be to the disadvantage of the interviewer. In the attempt to get quick answers, you might overlook some vital information related to the question, or disregard an aspect of the problem. Be sure and listen to what is being said, and not just hear what you want to hear.

When designing a questionnaire, think about the purpose of the questions you are asking. Ask yourself, "What am I aiming to find out?" If you are clear about the purpose of the question, you will be more effective in phrasing it accordingly. For example, in reviewing our proposal's questionnaire, focus on the questions carefully, and see if you can figure out what kind of information each one tries to obtain. If you have trouble, imagine how your interviewee can feel with your questionnaire, even though the question might be perfectly clear to you.

As a research subject answering these questions, I would have several problems. In question 1, what does "happiness" have to do with having many doctors? This is the way the medical system is; everybody specializes. I think the student was really aiming to find out whether or not patients are pleased with the present medical system, which forces a patient to have more than one general doctor and to go from specialist to specialist. If the above reasoning illustrates the intention of the question, then it needs to be rephrased. Perhaps it might be expressed as follows:

> In the present medical system, a number of specialized doctors provide care for the patient, rather than one general-practice-oriented doctor who takes care of all medical problems. How do you feel about this type of medical system?

If the researcher wants to ask more questions related to the basic statement, a listing of questions may follow, addressing aspects of the statement. However, since this is a formal interview rather than an informal conversational one, each question must address one particular idea which is clear to the interviewee. For example, in question 2, "satisfaction" with one's doctor can mean a number of things; therefore, a more precise statement is required. In general, ambiguous terms should be avoided. Clarity and precision in question framing are crucial in obtaining information as well as in evaluating it. Here, it is important to note that the order in which questions occur, as follow-ups of one another, should not be haphazard. Unless you are using warm-up questions, do not waste time inserting irrelevant queries. Organization is pertinent to speech and thought, especially during formal interviews. You want to make sure your subject is on the right track, thinking only about relevant information. Know what it is you are doing; be constructive. And be patient, developing the competence necessary for being a good interviewer. Most people have potential; skill comes with time and experience.

Another important point about investigation in general is to define the specific limits of the topic being researched. This provides for consistent data. To prove this point, consider these examples from the proposal: question 8 from the questionnaire, and, from the second paragraph of the proposal itself, sentence 13.

In question 8, the key word is "quality," and its evaluation, as indicated by five choices, is significant in reaching a conclusion based on the data. However, the word *quality* itself is unclear because it lacks a referent—quality in relation to what? Consider the following hypothetical situation. Let us imagine a neighborhood without a free clinic accessible to its members and whose residents are too poor to be able to afford personalized treatment. Obviously, having a clinic in this neighborhood is better than having no health care at all. So, from this perspective, if a clinic opened (where there had been no health care before) its quality would most likely be rated "excellent." However, if the clinic's supplies were rejects from other neighborhood clinics or pharmaceutical manufacturers, if the staff consisted of individuals who were fired from other medical institutions, disbarred, or were practicing without licenses, the quality of care would actually be questionable. In order to maintain uniformity in the data's meaning, pivotal words (those having a turning-point effect) must

1. be clearly defined by the researcher
2. be clearly defined by the user (in terms of the word's semantic value)

3. have the same referential value to all members of the group using them (so that a group's sentiments are projected correctly).

And, in order for point 3 to come about, there must be consistency between social variables which, when correlated with one another, indicate the problem as it is perceived by members of a cultural group. That is, simply to ask question 8 to persons belonging to the same culture, such as American culture, is not enough. But, to ask that question to twenty members of a particular ethnic group within the same neighborhood, with approximately the same educational background, pursuing similar jobs/careers/professions, participating in the same kinds of social activities, and having comparable financial status, then comparing the results with those based on the same number of individuals of another ethnic group, keeping variables constant, can produce a significant finding. Statements substantiated by qualitative or quantitative data are results of scholarly research. They are results of a tested hypothesis. They are reliable sources of information. Thus, they have become factual statements.

Participant-observation and ethnographic interviewing are not exclusive options; both are a part of anthropological research. However, in case of impossible restrictions, one method might take precedence over the other. It is always best to be flexible and resourceful when faced with unusual situations. This way you can be creative, using multiple approaches in incorporating the unexpected into your research in a constructive way. Regardless of the circumstances, if one follows sound research procedures and makes a commitment to his or her project, the opportunities for studying behavior, from an anthropological perspective, are infinite.

Organization of Research

By organizing your data collection, thinking about your data as you gather them, and interpreting their meaning during fieldwork, you will be able to write up your findings without much difficulty. Make sure you get an early start. Figure out how much time you have to work with, and divide it up into three sections. During the first of your three "chunks" of time, think about your general, and then your specific, topic, then consult your project director. Write up a proposal and review it with your project director. Concentrate on data collection and evaluation during the second phase. (This section should be longer in time than the other two.) Then use the last section for writing. Allow yourself ample time for unexpected demands associated with writing your final draft. Each individual has a unique writing temperament; be realistic about yours.

During data collection, equip yourself with research tools as required by your work. Whatever they might be, be sure they provide you with organization that makes your data comprehensible. It is advisable to keep a data book that records pure data, separate from a journal that records your commentary. This way your analysis will be kept separate from your description (see the appendix).

Besides the data book and journal, students may devise their own system of data recording. Some make check-charts that indicate patterns of correlated variables; others keep their findings in folders, arranged by cultural themes. Still others keep their findings in separate files by subject. Organization can also be made by categories of events. Whatever your system is, make sure its meaning is clear to you, so that after compiling your data, when you are trying to relate your findings to one another, they can be made to do so with ease.

While doing research one often discovers that certain literature has to be consulted. There is a constant feedback between data collection, thought, discussion, literature review, and reexamination. The process repeats itself, and with each new cycle an advance is made toward a deeper analysis.

Writing

When you are ready to write, make sure you have made sense out of your data so that all you have to do is translate them into print. Also, outline your paper or make a first draft so that its organization is clearly the best presentation of your work. Often a dramatic style that leads to climactic anticipation before introducing your key points will have a potent effect. Pay attention to technical details and keep interpretive information apart from data being presented. *Make sure the content reflects an anthropological perspective*. That is, make sure the concepts and the mode of analysis are anthropological. Remember, you have put time and effort into your creation; do not be apologetic, do not be on the defensive. You have learned about a social problem in depth; you have learned to do anthropological fieldwork; you have developed research skills. Now share your results with your reader.

Checklist

The following checklist summarizes the main steps in project planning, research, and presentation:

I. Choosing a topic
  A. Select a subject that interests you and is accessible.
  B. Narrow the field.
  C. As a result of narrowing, propose a testable hypothesis.

II. Designing the research project
  A. Define the population to be studied.
  B. Decide how to obtain a sample representative of the population.
  C. Choose appropriate methods of obtaining data, either
    1. participant observation,
    2. interviewing (determine appropriate questioning procedures), or
    3. other methods (e.g., simple observation, counting, etc.).
  D. Organize and record the data.

III. Writing the proposal (project design)
  A. State hypotheses—what you expect to find out.
  B. Justify the project. Answer reader's question: Why bother with this? Indicate the importance of your project to anthropology and general knowledge.
  C. State how you plan to go about doing the research. Be mindful of considerations of ethics and informants' privacy.

IV. Collecting the data (use and, if necessary, modify the intended procedures)

V. Evaluating the data
  A. Organize quantitative data into tables for presentation.
  B. Decide whether data support hypothesis. If not, think about why.
  C. Identify and explain findings you did not expect.
    1. Are the hypotheses not valid?
    2. Are there problems with research design?
    3. Are there other reasons?

VI. Writing the final report
  A. Clearly present the points made in the checklist.
  B. Give a bibliography following AA style (used in this book).
  C. Do a self-analysis (see the essays by Faber and Larson).

Appendix

The following unedited excerpts illustrate what is meant by keeping a data book and journal respectively. These examples are taken from my own fieldwork.

DATABOOK: Male-Female Relationships among ______ Ethnic Students

DATE: Oct. 23, 1975
Place: School Cafeteria
Time: 12:00–1:30 P.M.
Observation: Conversation around lunch table
Members: 15 indiv. / 8 males—7 females
Description: (third week in a row, same composition as always)

3 sq. tables put together—long rectangular one / same location in caf.—immediately to the right of entrance / males cluster together at head of table (close to door), ⅔ space occupied / females at end of table (in relation to males), ⅓ space occupied / males dress casually (blue jeans, shirt, sweater) / females dress fashionably (tailored suits, skirts with fancy tops, boots with heels, etc.) / males fairly groomed (½ shaven, ½ not; ½ hair combed, ½ not and dirty) / females very well groomed (clean hair—styled, make-up / males talk to one another about politics, read the newspaper, smoke, help each other with engineering problems, etc. / females talk among themselves with low tone, discuss fashion, gossip, community events, some classes / Individuals come and go because of lunchtime but same composition of group and same behavior patterns persist.

JOURNAL:

Today marks the 11th time I've come to the same place, same time, same people, no longer coincidence, there is definite pattern as displayed by seating arrangement and restricted verbal intercourse. It seems that in public, topic-to-be-discussed-by-gender is culturally defined and mode of communication in terms of conveying messages that are nonverbal. Females dress and act and talk in a way which makes them "feminine" and men "masculine." It is more important that the female be "presentable" because males have prerogative of choice. (They take initiative.) But females let out vibes as to who they want. They use a lot of eye contact as a means of flirting. Repeated, quick glances until contact is established and one long glare (discreetly) between the couple. I was trying to talk to ______ but she only payed attention with her ears, so I looked up quickly to the male section and saw ______ looking back at her over his newspaper. Both knew I "caught" them and they blushed. I had exposed their indiscretion.

A lot has to be analyzed about circumstantial manipulation and distribution of power by gender. And how individuals are treated when they break taboos. No doubt there is contradiction between propagandized ideology that gets the group together and that of actual behavior. Of course—why does it occur and what becomes of the role of ideology?

Note: Look up references—Body Language and Symbolic Communication ______.
Ethnic Group's Cultural Taboos / Political Ideology of the ______.

Lebriz Tosuner-Fikes's guide (essay 1) has provided detailed discussion of terms and techniques useful in planning and carrying out research. Here Linda Place offers briefer guidance, in a more outlined form. Students should pay attention to this brief guide in planning and preparing their research reports. Christina Hill's paper (later in this volume) is a good example of the usefulness of this brief guide in organizing the research and its presentation.

2 / Brief Guide to Presentation of Field Research

Linda Place

The following sections should be discernible in your research report. It is not necessary that they be indicated by subtitles (in fact, it is rather unusual to see such formality in an anthropological report, although there is nothing wrong with this approach); however, it is essential that all sections be included in the specified order.

Introduction and Statement of Hypothesis

The introduction should provide information on the general topic selected for research. Why was it chosen and what is its relevance to anthropology? Your hypothesis statement should indicate the more specific aspect(s) of this particular research. This is a statement of the problem(s) or question(s) under investigation. By definition, a hypothesis should be testable. The testing of your hypothesis is the goal of this project. (The posing of a series of questions and the search for viable answers to these questions is essentially the same procedure.) This is a very important stage of research, and extreme care should be given to its exposition. Identify all sources that have inspired or influenced your hypothesis selection.

Operationalization of Hypothesis

Since the object of research is to prove or disprove one's hypothesis (or to find answers to one's questions), it is necessary to delineate methods and techniques appropriate to such a task. Your first concern should be with making explicit the terminology of the project. Anything that might lend itself to ambiguous or multiple interpretations must be defined precisely, in order to make the researcher's usage apparent. Categories of evaluation are especially subject to misinterpretation; thus, great care should be taken in their definitions. For example, in terms of personal space, does *close* mean an inch or a foot apart?

It is also of value to define your population. What are its boundaries? Is your sample representative of the population? Will the limits of your sample affect your ability to generalize your results to American culture as a whole? Such factors as age, sex, class, occupation, and ethnicity that are relevant to the research should be incorporated in your definition.

The next item of concern is the collection and recording of your data. How did you go about answering the research question? Did you use a participant-observation approach, interviews, questionnaires, or some combination of approaches? What was the frequency and total time period of the data-collecting situation? What were the relevant conditions of the data-collecting situations, such as type of environment(s)? How were data recorded? (Ideally, these concerns will have been addressed prior to submitting a prospectus.)

Careful attention should be paid to detail in these first two sections of your presentation. The less you leave to the imagination of the reader, the stronger your arguments will be. Being explicit about your assumptions, definitions, and approach contributes to an overall impression of research integrity.

Presentation of Data (Results)

There are two aspects of this section which, depending on the particular project, may be handled separately or together. These are: (1) a strict statement of the raw data and their analysis, and (2) a discussion of the data or the results based on data analysis. If your project had a good numerical base suggesting a statistical type of analysis, then these aspects will probably be incorporated separately into your report (e.g., in the form of one or more tables). If, on the other hand, your project focuses on more generalized statements that are not reducible to discrete units, analyzed data may be presented in conjunction with its discussion. In either case, the material presented in this section will be regarded as the foundation for all conclusions reached as a result of your project.

Anomalies in your data should not be disregarded. Instead, they should be included in your presentation along with possible explanations for their occurrence. If no pattern reveals itself after the data are analyzed, then this should be stated and dealt with. Be certain, however, not to restrict yourself to searching for patterns that support your hypothesis. Other patterns may be present that you hadn't anticipated.

Summary and Concluding Statements

A brief summary that ties together the salient features of the project may be of value at this point. This should function only as a means of presenting a coherent statement of what has already been stated at length.

By far the most important section of this project report is your conclusion. Was the hypothesis supported? If so, what substantiated its proof? Is it likely to hold true under similar conditions elsewhere?

If your hypothesis was not supported by your research, give reasons for this lack of support. Questions you should ask are of the type: Were there sufficient data? Were observations made objectively or is researcher bias distorting results? Were research conditions conducive to good data collection? Was the hypothesis valid? What other questions are suggested by these results?

These questions regarding hypothesis proof are only a beginning. Whether or not your results are positive with respect to your original thesis, the contribution of your research to anthropological concerns should be the emphasis of your conclusions. What has your research revealed about contemporary American culture? It would be of value to refer to similar conceptualizations or correlations found in lectures or reading assignments, indicating a firm understanding of the more general issues involved.

Bibliography

Follow this volume for the proper form of in-text citation and bibliographic entry.

Unifying Themes in Contemporary American Culture: Analysis of the Mass Media by Etic, Structural, Symbolic, and Neo-Freudian Techniques

Essays 3 through 10 focus on unifying forces in contemporary American culture, particularly on the mass media. Various anthropological research strategies and modes of analysis are defined and applied to American "popular culture." Native anthropologists are anthropologists who study their own culture; they must take particular care to avoid their own emic biases—the opinions, viewpoints, and interpretations they have learned through a lifetime of enculturation. However, anthropological training and cross-cultural experiences provide native anthropologists with a measure of detachment that can reduce emic biases and help them develop an etic, or more impartial observer's, perspective on behavior in their culture. Along with structural, symbolic, and neo-Freudian analysis, the contrast between emic and etic perspectives is illustrated through consideration of football, rock music, amusement parks, fast-food restaurants, films, fairy tales, myths, advertising, and adolescent pulp literature. Other ways of analyzing the mass media, including quantitative and statistical methods, are illustrated in later essays, particularly essays 22 through 24. Most of the selections in Part 2 are by professionals.

This updated and slightly revised version of a chapter of my textbook *Anthropology: The Exploration of Human Diversity* (2nd ed.) introduces and illustrates some of the techniques of anthropological analysis that are used in the student papers. The approaches emphasized in this essay are emic and etic research strategies, symbolic and structural analysis, and the study of ritual. Each method is applied to an aspect of contemporary American mass culture, so that students can see that their shared experiences, attributable to common enculturative forces, particularly the mass media, are amenable to anthropological analysis. Essays 4 through 9 are additional illustrations of ways in which techniques originally designed for the study of simpler societies can be extended to our own.

3 / Anthropological Analysis of Mass Enculturation

Conrad Phillip Kottak

It is appropriate that anthropology, traditionally the study of distant, small-scale, "exotic" societies, should also study American society and culture. Anthropology aims, after all, at being a science of human behavior that is concerned with social and cultural universals, generalities, and uniqueness. American culture is a particular cultural variant, as interesting, as exotic, and as unique as any other. Americans are similar to members of other cultures in some ways, different in others. Anthropological techniques developed in smaller-scale societies, where sociocultural uniformity is more marked, can nonetheless contribute to an understanding of American life.

In describing and analyzing American culture and society, American anthropologists enjoy an advantage. Although their training in anthropology and their familiarity with other cultures grant them a certain degree of removal and objectivity in studying American culture that most American natives and other scholars lack, American anthropologists *are* Americans. Anthropologists from various third world nations have correctly asserted that their life experiences as natives, combined with their scientific training and objectivity, give them an added advantage in anthropological studies in their own countries. The same applies to American anthropologists studying American culture. As native anthropologists we are frequently full participants as well as observers, often emotionally and intellectually caught up in the events and beliefs we are describing.

One irony of the historical expectations of anthropology is striking. Anthropologists write extensively about foreign areas in which they have spent several years doing fieldwork. Yet is it not even more appropriate for them to use their knowledge of anthropology to try to comprehend a culture they have lived in and observed for a lifetime—a good part of it as an anthropologist? The examples in this essay demonstrate that I and several of my colleagues deem such native anthropology a challenge and a pleasure.

However, native anthropologists must be particularly careful to resist their own emic biases (their prejudices as natives) and to be as objective in describing their own cultures as they are in analyzing others. Any anthropologist must also be aware that natives often see and explain their behavior very differently from the way the anthropologist does. In most of the examples that follow, the etic perspective is taken. Most Americans have probably never considered the possibility that such apparently secular, commercial, and recreational institutions as football, rock music, Walt Disney Enterprises, and fast-food restaurants have things in common with religious beliefs, symbols, and behavior in our own and other societies. Most Americans have probably thought little about why so many young Americans simultaneously appreciate such apparently different events as football games and rock concerts. Yet anthropological techniques usually used to deal with, say, myths of South American Indian tribes and rituals of cultivators in East Africa can show not only how football and rock are related, but how each relates to equally significant aspects of American ideology. Similarly, techniques developed to analyze rituals and myths in nonindustrial societies can enlighten us about "Star Trek" and its devoted following, science fiction, the Disney organization's contributions to American culture, and our behavior at what most Americans consider the most ordinary places we frequent—fast-food restaurants.

Anthropology textbooks often stress the discipline's value for understanding ourselves; by studying other cultures, they say, correctly, we learn to appreciate, understand, and question our own. However, since only recently has the United States become an area significant to anthropologists, this assertion usually remains undemonstrated. In the following examples it is shown that the very same techniques that anthropologists use in describing and analyzing other cultures can be applied to our own. This essay will also raise another intriguing possibility. By examining whatever resistance you as an American native may have to accepting these anthropological analyses of your own beliefs, values, and behavior, you may come to see that any natives, in any culture, may have similar reactions to an anthropologist's account of their culture.

Furthermore, Americans may not find the arguments in this essay convincing. In part this may be because you are a native and know so much more about your own culture than any other. But in part it may also be because in trying to extract culture (*shared* programs for behavior) from a variety of individual opinions, acts, and experiences, we, as anthropologists, sometimes depart from areas that can be quantified—such as household composition, population density, poverty, wealth, and socioeconomic variation in general—and enter a more impressionistic domain, one in which analysis sometimes seems as akin to philosophy or to the humanities as to science. Certainly, you will be right in questioning some of the conclusions set forth here. Some are surely debatable, some perhaps just plain wrong. But if they illustrate how anthropology can be used to shed light on aspects of your own life and experience, and used to revise and broaden your understanding of your own culture, they will have served a useful function.

A final word about culture, ethnocentrism, and native anthropologists is needed on the application of anthropological techniques to American culture. For anthropologists, *culture* means much more than refinement, cultivation, education, and appreciation of the fine arts—its popular usage. Curiously, however, when some anthropologists confront their own culture, they seem to forget this. They carry an image of themselves as adventurous and broad-minded specialists in the unusual, the ethnic, and the exotic and, like other academics and intellectuals, tend to regard aspects of American "pop" culture as trivial and unworthy of serious analysis. In doing so, they demonstrate their own ethnocentrism as American natives and reveal a bias that comes with being members of the academic-intellectual American subculture.

In examining American culture, native anthropologists must be particularly careful to overcome this subcultural bias. That twenty million Americans, mostly women, watch soap operas daily is surely a highly significant sociocultural fact of American life. That seventy-nine million people tuned their televisions to the first annual Super Bowl game (Arens 1976) is almost as impressive and as important in understanding American culture as the fact that televisions outnumber toilets in American households. My own research on Michigan college students may be generalizable to other young Americans. They visit McDonald's more often than houses of worship. No more than five percent of them had *never* seen a Walt Disney movie. In each case, the number of University of Michigan college students questioned who had never entered a Protestant, Catholic, or Jewish house of worship far exceeded the number who had never eaten at McDonald's, seen an episode of "Star Trek," or attended a rock concert. If true of young Americans generally, as I suspect they are, these highly significant facts about Americans and their culture suggest major twentieth-century modification in American enculturation patterns. Certainly, any extraterrestrial anthropologist doing fieldwork in the United States would stress them. They represent major, perhaps dominant, themes in contemporary American culture. Surely, they merit anthropological study.

The cultural phenomena examined in this essay (football, rock music, Disney products, and fast-food restaurants) are all manifestations of contemporary mass culture. That is, they involve settings, beliefs, and behavior patterns familiar to most Americans, rather than the distinctive customs, beliefs, or behavior of particular regions, classes, or ethnic groups. Sociocultural diversity in the contemporary United States is the subject of several of the essays that come later in this volume. Cultural unity will be the focus here.

Now that we have looked at some of the general problems associated with any anthropological study of American culture, some background information about several aspects of ritual is needed. In American culture, certain events, beliefs, and activities, on the surface, appear ordinary and mundane, even profane. However, these components of American culture have some functions and aspects analogous to religious rites and doctrines in our own and other cultures.

Rites of Passage

Early in this century Arnold van Gennep (1960, 1st ed. 1909), a Belgian anthropologist, studied rites of passage in a variety of societies. Passage rites are encountered in every society; they are exemplified by such phenomena as vision quests of certain Native American populations in North America. As boys moved from boyhood to socially recognized manhood, they temporarily separated themselves from their communities to journey alone to the wilderness. After a period of isolation, often accompanied by fasting and drug consumption, the young men would see a vision, which would become their personal guardian spirit. On return to their communities they would be reintegrated as adults. In contemporary societies, rites of passage include confirmations, baptisms, bar mitzvahs, and fraternity hazing. Passage rites do not refer only to such changes in social status as from boyhood to manhood, or from nonmember to fraternity brother, but apply more generally to any change in place, condition, social position, or age. Examining data from a variety of societies, van Gennep generalized that all rites of passage have three phases: separation, margin, and aggregation. Separation is exemplified by the initial detachment of individuals from the group or their initial movement from one place to another; aggregation, by their reentry into society after completion of the rite. More recently, anthropologist Victor Turner (1974) has focused on the marginal period, the position between states, the limbo during which individuals have left one place or state but have not yet entered or joined the next. Van Gennep (1960) used the Latin term *limen* (threshold) to refer to this in-between period, and Turner's designation of it as the *liminal* phase of a passage rite will be used here.

On the basis of data from several societies, Turner (1974) identified generalized attributes of liminality. Liminal individuals occupy ambiguous social positions. They exist apart from the status distinctions and expectations of normal social life, living in a time out of time. They are cut off from normal social intercourse. In contrast to the vision quest, which is individualistic, passage rites are often collective. A group of people—boys undergoing circumcision, fraternity initiates, men attending military boot camps, football players at summer training camps, women becoming nuns—pass through the rites together. Turner points out that liminal periods are ritually demarcated by a variety of contrasts with normal social life. Among the Ndembu of Zambia, whom Turner studied, a newly chosen chief had to undergo a passage rite before taking office. During the liminal period, his past and future positions in society were ignored, even reversed, and he was subjected to a variety of insults, harangues, instructions, and humiliations.

Turner lists a number of contrasts or oppositions between liminality and normal social life (see table 1). Most notable is the social aspect of collective liminality that he calls *communitas*. People who experience liminality together characteristically form an egalitarian community; whatever social distinctions have existed before, or will exist afterward, are temporarily forgotten. Liminal individuals experience the same treatment and conditions and are expected to act alike. Liminality may be marked ritually and symbolically by reversals of ordinary behavior. Sexual taboos may be intensified or, conversely, sexual excess may be encouraged.

Turner points out that not only is liminality always a temporary

TABLE 1 Oppositions between Liminality and Normal Social Life

Liminality	Normal Social Structure
Transition	State
Homogeneity	Heterogeneity
Communitas	Structure
Equality	Inequality
Anonymity	Names
Absence of property	Property
Absence of status	Status
Nakedness or uniform dress	Dress distinctions
Sexual continence or excess	Sexuality
Minimization of sex distinctions	Maximization of sex distinctions
Absence of rank	Rank
Humility	Pride
Disregard of personal appearance	Care for personal appearance
Unselfishness	Selfishness
Total obedience	Obedience only to superior rank
Sacredness	Secularity
Sacred instruction	Technical knowledge
Silence	Speech
Simplicity	Complexity
Acceptance of pain and suffering	Avoidance of pain and suffering

Source: Adapted from Victor W. Turner, *The Ritual Process.* Copyright © 1969 by Victor W. Turner. By permission of Aldine Publishing Co., Chicago.

part of any passage rite, it may, in certain social contexts, become a permanent attribute of particular groups. This will occur, Turner suggests, in the most socially diverse societies, presumably state-organized societies and particularly modern nations. Thus, within nations, religious sects often use liminal characteristics to set themselves off from the rest of the society. Such requirements as humility, poverty, equality, obedience, sexual abstinence, and silence may be conditions of sect membership. The ritual aspect of persons, settings, and events may also be communicated through liminal attributes that set them off as extraordinary—outside normal social space and regular time. Turner's examination of liminality contributes to a useful framework for examining many ritual and quasi-ritual aspects of American culture, including those associated with football.

Football

Football, Americans say, is only a game. Yet it has become one of the most popular spectator sports in the United States. From August to

January, from Friday to Monday, Americans can follow football. In places as diverse as Ann Arbor, Michigan, and Los Angeles, California, people spend Saturday traveling to and from, and attending, college football games. Smaller congregations meet for high school games. Vast audiences watch televised football; indeed, more than half the adult population of the United States regularly watches the annual Super Bowl. As anthropologist William Arens (1976) points out, football is clearly not simply a preoccupation of "Middle America," but of all America, and as such is a major aspect of American culture. An interest in football unites Americans regardless of ethnic group, region, state, urban, suburban, or rural residence, religion, political party, job, status, wealth, gender, or sexual preference.

While football originally may have become popular as a vehicle of college spirit, the popularity of contemporary college football and particularly of professional football depends directly on the mass media, especially television. Arens (1976) analyzes reasons for this popularity. Questions often asked are: Is football, with its territorial incursion, hard hitting, and physical violence—occasionally resulting in injury to players—popular because Americans are a violent people? Are Americans naturally bloodthirsty? Are football spectators vicariously realizing their own hostile, violent, and aggressive tendencies? Arens discounts this approach. He points out that football is an almost uniquely American pastime. Although a similar kind of football is played in Canada, it is considerably less popular than in the United States. Baseball has become a popular sport in the Caribbean, some areas of Latin America, and in Japan. The popularity of basketball is also spreading. And throughout most of the world, soccer is the most popular sport. Arens argues that if football were a particularly effective channel for expressing aggression, it would have spread to many other countries, where people have as many aggressive tendencies and hostile feelings as Americans do. Furthermore, he suggests that if a sport's popularity really rested on a bloodthirsty temperament, boxing, a far bloodier sport, would be America's national pastime. Arens concludes that the explanation lies elsewhere, and I agree.

Arens contends that of all the sports, football best represents, or symbolizes in a public context, certain major features of American life. In particular, it echoes group coordination through elaborate variation, specialization, and division of labor—pervasive features of American society and economy. Susan Montague and Robert Morais (1976) take his analysis one step further. They argue that Americans appreciate football because they can recognize in it a miniaturized and simplified version both of the structure of modern industrial organizations and of the behavior deemed appropriate by such business enterprises, and they can master it as such. Bureaucracies, whether in business, universities, or government, are indeed perplexing and mysterious. Ordinary workers, faculty members, and students find it difficult to understand the structure of such organizations as businesses and universities, and to comprehend how decisions are made and rewards allocated.

Montague and Morais link values extolled in football to those associated with success in business. Ideal characteristics of football players, as of success-oriented businesspeople, include hard work, diligence, dedication, and denial of individual self-interest for the good of the team ("teamwork"). But, the anthropologists argue, rewards for such behavior in business are not always forthcoming. Because of the complicated and often capricious nature of decision making within large organizations, workers are not always assured that they will be rewarded for their dedication and good work. For precisely this reason football is popular. Any fan can, through careful study and observation, become an expert on the rules, teams, scores, individual statistics, and patterns of play in football.

Even more important, football demonstrates to fans that the values stressed by business really do pay off. Teams whose players and coaches work hardest, show the most team spirit, and best develop and coordinate the talents of their players can be expected to win more often than others. Both teams and players receive large bonuses. Classic capitalist values, which are still presented as guides for success in American business, are represented by and affirmed through football. Football is popular therefore because it is so well suited to the American economy, society, and traditional values.

Football and Rock Music

Many young Americans are equally devoted to football and to rock music. How can this common interest in such diverse areas be explained? Montague and Morais (1976) attribute the popularity of football and rock, often among the same people, to the association of each phenomenon with an opposite but equally important theme of American life. Football is associated with the *technical* side of American society and the American economy; it provides a simplified model of American industrial organizations and illustrates that business values and appropriate capitalist behavior are rewarded in the end. Rock, on the other hand, can be linked to a long-existing but more recently articulated theme in American culture—the desirability of *creative* expression, the value of doing one's "own thing." Whereas in football, teamwork, diligence, talent, and self-denial are rewarded, in rock, creative expression is rewarded, not only materially (a fact that many rock stars try to play down), but also in the communitas, communion, even love, between audience and star.

Rock and football can coexist in American society and be appreciated by the same people because, despite their contrasting nature, both are essentially "as American as apple pie." The popularity of both rests on changes in the mass media in the mid-twentieth century. Both are totally compatible with the individualistic focus of American society, with the decline of kinship values and a shift to social interactions and associations with nonrelated individuals. Both exalt individualistic values: the individual football player, along with the team, is rewarded for working well with others with money, fame, and fans' appreciation; the individual rock star is rewarded with fans' affection and devotion as well as with wealth and renown.

Rock music and football, although opposites in several concrete ways, are also linked, since both depend on and express common themes and values in contemporary American culture. Consider a series of contrasts or oppositions that, by making rock and football seem like totally different institutions, mask in our minds their common relationship to individualism and other American values. They involve behavior of performers and audiences; symbolism associated with performances and games; and ritual aspects, including liminality. The analysis of similarities and differences between rock and football that follows exemplifies *structural analysis*, a technique of comparison that has recently found favor among American anthropologists.

Structural Analysis

During the 1970s, Claude Lévi-Strauss, a prolific French anthropologist, has become a major figure in both European and American anthropology. First in *Structural Anthropology* (1967), and subsequently in several other books, Lévi-Strauss has advocated an approach to social and cultural data known as the structural method of analysis. Lévi-Strauss's *structuralism* rests on his assumption that the human mind has certain universal characteristics and that they reflect common structures of the human brain. Accordingly, he believes that similarities in the brain structures of people everywhere lead them to think similarly no matter what their society or cultural background. Among the universal characteristics are the need to classify—to impose order and arrangement on aspects of nature, on people's relationship to nature, and on the relationships between people. According to Lévi-Strauss, a universal aspect of the need to classify is opposition or contrast. Although many, perhaps most, phenomena are continuous rather than discrete, Lévi-Strauss argues that the mind, because of its need to impose order, makes them more discrete and different than they really are. Things in nature that are continuous, that are quantitatively rather than qualitatively different, are made by classification to seem absolutely different and discrete. Scientific classification is merely a western academic expression of the universal need to impose order by classifying. One of the most common means of classifying is by *binary opposition.* Good and evil, white and black, old and young, high and low are oppositions that, according to Lévi-Strauss, reflect the human need to convert continuous contrasts into absolute contrasts.

Binary Oppositions between Rock and Football

In contemporary American culture, as Montague and Morais (1976) have pointed out, several binary oppositions differentiate rock and football. Yet rock and football are also similar because they exaggerate, although in different directions, the *same* set of prominent themes in contemporary American culture. Rock and football are associated with polar ends of several contrasts significant in American life. By converting differences of degree into differences of kind—binary oppositions—rock and football set themselves off from each other and from ordinary life. Let us consider some of these binary oppositions. (See table 2.)

Individual versus group. We all do some things alone, others in groups. We are neither total loners nor totally social creatures. However, the contrast between individual and group is converted into a binary opposition, like other differences that differentiate rock and football. Rock musicians perform individually or in small groups, whereas a football team is a coordinated, highly specialized, tightly integrated group. Football players are regularly replaced during a game; substitutions are allowed. Rock focuses on the individual, even in small groups. The Beatles, for example, began as a group of individual artists and writers; they became a group but when each decided to follow his own career, the group dissolved because substitutions were impossible.

TABLE 2 Binary Oppositions Represented in Rock Concerts and Football Games

Rock	Football
Individual or small-group activity; focus on individual performer even in small groups	Coordinated, highly specialized, tightly integrated group activity
Expresses individualistic, creative theme of American culture	Stresses technoeconomic basis of American industrial society
Deemphasizes "masculinity" of males and "femininity" of females	Exaggerates "masculinity" of males and "femininity" of females
Deemphasis of sexual differences permits heterosexual acts to be mimicked or imitated in performances	Exaggeration of sexual contrasts permits quasi-homosexual behavior on the field
Stresses performer's illicit behavior; sexual license; promiscuity expected	Stresses performer's clean living; sexual prohibitions; abstinence expected
Audience participates in performance, interacts with performer	Audience observes performance, rigidly separated from performers
Performers allowed to consume drugs and alcohol, to behave obscenely	Audience permitted to consume drugs and alcohol, to behave obscenely

Creative versus technical. All of us sometimes "do our own thing," sometimes carefully mesh our activities with those of others. But the creative-technical distinction—one of degree—has also become a binary opposition separating rock and football. Rock stresses the individual artist and exemplifies the American value of individual fulfillment through creative expression. Football involves coordination, teamwork, and self-denial, and the submergence of the individual in the social; it reflects the values of technique, endurance, and skill in American society and economy.

Deemphasis of gender differences versus exaggeration of gender differences. There are contrasts in the typical personalities, dress, and behavior of American males and females, but they are quantitative rather than qualitative. Rock and football deny this, but in opposite ways. Rock tends to deny gender differences, while football exaggerates them. The dress, posture, and gestures of some male rock stars often disguise the fact that they are males. Female rock performers also depart from traditional American feminine cultural norms; sometimes this is expressed in dress, but it is more obvious in behavior, including loudness and exuberance. Arens (1976) has pointed out that the uniforms of football players exaggerate male anatomy. Shoulder pads are the cultural equivalent of the shoulder manes of adult male baboons. Skintight trousers and metal codpieces complete the supermasculine effect. As much a part of the football ritual as the game itself are the ultrafeminine cheerleaders and "pom-pom girls," who are selected more because they meet American standards of beauty and appropriate feminine personality than because they are effective in rousing fan support. An obligatory aspect of a football telecast is the camera pan of attractive young women spectators. In football the differences between male and female are celebrated symbolically; in rock they are hidden or denied.

Heterosexual versus homosexual. By denying sexual differences, rock allows performers to include sexually explicit material in their songs and gestures. A variety of sex acts are imitated, mimicked, and simulated on the stage (Montague and Morais 1976). In contrast, because football players exhibit such exaggerated masculinity, their hand holding, hugging, and bottom patting, which in other contexts might suggest homosexual behavior, are socially acceptable.

Illicit versus licit. In another binary opposition, rock and football create stereotypes of socially disapproved and approved behavior. Football players, particularly college players, are supposed to exemplify "clean living"; this is an aspect of their self-denial. Smoking, drinking, drug consumption, and gambling are taboo (Montague and Morais 1976). Arens (1976) has pointed out similarities between sexual taboos associated with football and ritual prohibitions against sexual relations in other societies. In many societies, men are expected to be sexually continent before a hunt or a raid; even sexual relations with their wives are tabooed for a period established by convention. Football players observe similar sexual taboos. During summer training camp—a liminal period prior to the start of the football season—professional players are isolated from their wives or other women. Both college and professional players are also expected to abstain from sex on the night before a game. Like liminal figures generally, they observe a variety of taboos on eating particular foods, smoking, and drinking. At the University of Michigan, football players, coaches, and staff spend the Friday nights before home games in a local hotel. They read, participate in prayer groups, and watch such movies as *Patton*, which provide lessons in the values associated with football: discipline, teamwork, coordination, dedication, and submission to authority.

A rock star's life-style is touted as just the opposite of that of the football player. Rock stars are expected to consume a variety of drugs and to be sexually promiscuous. The teenage groupie, available for sex after performances, fills a functionally specialized role within this set of expectations. Expression through drugs, sexual excess, and other behavior condemned by the larger society is an aspect of the individualistic, creative focus of rock. Whereas football players symbolize their submersion in society and submission to authority by watching *Patton* and praying together, rock performers use drugs and sex to symbolize their transcendence of society and freedom to do their own thing.

Incorporation versus separation. Both football games and rock performances share with many religious rituals certain aspects of liminality—namely, in being set off from ordinary time and space—but again they do it in different ways. This aspect of football is particularly obvious. Football games normally take place in special-purpose structures intended primarily or exclusively for sports events. Home games are scheduled at more or less invariant hours on specific days of the week. Football players' liminality is not limited to their separation from ordinary social life the night before the game, but lasts until the game is over. Audience and players are physically and ritually separated in the football stadium. Barriers of grass or astroturf clearly demarcate the field from surrounding stands, and the actual playing surface is marked off by lines. Total separation of audience and players is expected. Television commentators serve as guardians of the game's morality, scolding fans who throw paper, smoke bombs, and bottles onto the field. Nets behind goal posts protect the game's most sacred object from defilement by contact with the audience. Only at game's end, and particularly after a very important game or at the conclusion of the season, are fans allowed on the field. The separation of players and spectators finally ends as fans congratulate the players; the end of the separation is symbolized by tearing down the goalposts after key games.

In contrast to football, where players are temporarily isolated from normal social life at specified times of the year and days of the week, in rock, performers are more permanently isolated from society by the expectation of individualism and excess in their life-styles. Yet the separation of spectators and performers in football is inverted in the rock concert, which, if successful, becomes a true communion of performer and audience (Montague and Morais 1976). Ideally, the audience and rock performer become one; the performer is rewarded in love and community for creative achievements. Stars sometimes move into the audience, touching their fans and letting their fans touch them. The emotions generated during a rock concert are more sensual, and through the performer's instigation, more sexual than during a football game.

Performers' excess versus audience's excess. In football—at least in college football—players and fans are expected to show opposite types of behavior. For the players the game is a time to demonstrate that clean living translates into victory, but for spectators it is conventionally a time of excess, expressed through continuous ingestion of consciousness-altering substances—mostly alcohol and marijuana. Turner (1974) reports that among the Ndembu of Zambia, customary parts of certain rituals are collective expressions of obscenity. So, too, at University of Michigan football games, spectators are allowed to violate normal taboos against certain words, shouting in mass "Bullshit" when the home team fails to gain necessary yardage. The contrast between the expressive behavior of performer and audience is less marked in a rock concert, aiding the communion that is supposed to develop. Still, stars often consume drugs as part of their performance. Although spectators may smoke or drink, especially at outdoor concerts, they have come to be "turned on" more by the performance than by drugs.

Broader Themes of Opposition in American Life

Lévi-Strauss has shown how specific oppositions contained in the myths and social structures of nonindustrial peoples can be linked to larger-scale oppositions that reverberate throughout their lives and cultures. Consider briefly now how the specific structural analysis of rock and football just given might be enlarged to reveal broader themes of opposition in American life.

We might say, for example, that rock is to football as night is to day. This opposition is derived from a number of material and symbolic aspects. Football games are usually played—at least in colder areas of the United States—during the day; rock concerts are usually held at night. Americans symbolically associate day with cleanliness and propriety, night with impropriety, illicit activity, and murkiness. Or we might say that football is to rock as cold is to hot. Football is a cold-season game, whereas rock concerts, particularly those held outdoors, are associated with summer and California. Rock—symbolically stressing affect, sexuality, vibrant colors, loud music, and psychedelic patterns—seems, in Marshall McLuhan's terms, a "hot" medium. Football—with its technical aspects, its exclusion of women from the game and of fans of both sexes from the playing field, and its limited, traditional, and clearly bounded field and uniform colors—suggests a "cold" medium, again in McLuhan's terms. Surely a rock concert comes closer to most Americans' image of hell than does football.

Rock is left and football is right. Americans classify political positions on a scale of left-right opposition. Republicans are supposed to represent right, Democrats left. Rock stars tend to support Democrats or positions to the left of Democrats; football players are thought more likely to support Republicans and to take right-of-center stands. Republican Richard Nixon cultivated ties with football coaches and used football terms to describe nonfootball situations. Republican Gerald Ford was a college football player. Republican Ronald Reagan's most famous movie role was as a football player. On the other side, during their presidential campaigns, Democrats Jimmy Carter and Jerry Brown both actively sought the public support of rock stars.

Thus the specific binary oppositions that differentiate rock and football link up with larger values and themes of opposition in American culture. Through association with opposite ends of the same contrasts, rock and football are set off from each other and from other domains of American life, granting to each its own special niche in our culture. Although rock and football symbolize different extremes, we should not lose sight of the similarities that explain their popularity and significance to contemporary Americans. Both rock and football exaggerate—differently—the same cultural themes. Both express the individualistic focus of contemporary American society. Both rock and football are big business, and the popularity of both rests on mid-twentieth-century changes in the mass media, as is true in the examples that follow.

As an American native, you probably question this structural anthropological perspective on football and rock, just as South American Indian informants might doubt a structural analysis of their important myths. Structural analysis aims not at *explaining* otherwise hidden relations, themes, and connections among aspects of culture, but at uncovering them. Structuralism is as akin to the humanities as to science.

In fact, structuralist techniques are now widely used in literature and art as well as in anthropology, illustrating anthropology's holism—its links to other disciplines. Any structural analysis is difficult to confirm. How one chooses which of two or more structural analyses is more accurate or more revealing usually reflects personal opinion. But after thinking about rock versus football and their relationships to American culture in terms of structural analysis, one can never again view rock as merely a form of entertainment or football as only a game. Structural analysis can, I believe, be enlightening as one among many tools available to the anthropologist.

Walt Disney as Creator: Walt Disney World as Shrine

American natives think of football as "only a game" and rock music as one among many forms of entertainment. Along the same lines, most also think of Walt Disney as a highly successful businessman whose commercial empire has been built on movies, television programs, and amusement parks. Again, however, etic analysis reveals that there is much more to Walt Disney Productions than most Americans have recognized. Specifically, our common exposure to Disney products provides a highly significant common enculturative experience for contemporary Americans, particularly for those born after 1937, when *Snow White and the Seven Dwarfs*, Disney's first full-length cartoon, was released. Furthermore, there is a quasi-religious aspect to our relationship to Disney and his works; this emerges most clearly in Americans' behavior at the two Disney amusement parks, Disneyland, in Anaheim, California, and Walt Disney World, near Orlando, Florida. The following account is intended to illustrate certain similarities between visits to these amusement parks and pilgrimages to such religious shrines as Mecca, the holy city of Islam; Lourdes in France; Fátima in Portugal; and, particularly, Bom Jesus da Lapa in the arid *sertão* of northeastern Brazil (see Gross 1971). Since I have never been to Disneyland, my observations are based on several trips—as participant and observer—to Walt Disney World.

The quasi-religious attributes of Walt Disney World rest on Americans' prior exposure to other, nationally available products of the Disney organization. Disney's influence on Americans has survived his death. His classic movies are regularly rereleased; noncartoon full-length films have been shown on national television; cartoon characters live on in short features, television programs, advertising, and—in the flesh—at the Disney amusement parks. Though the creator was mortal, his vision has posthumously guided the organization; indeed, he left his successors a plan. This scheme included future movies and other entertainment projects, as well as completion, to his specifications, of an obvious shrine to his memory—Walt Disney World. The willingness of Americans to journey hundreds or thousands of miles to Disneyland or Walt Disney World reflects decades of work in preprogramming our populace in Disney lore.

Particularly important in Disney lore are the images of childhood fantasy—the cartoon characters, often unusual humans and humanlike animals—that continue as very important components of the mythology of American childhood. A common vision of a witch is Snow White's stepmother, particularly as she holds a poisoned apple. *Snow White* again provides our image of dwarfs, Peter Pan's Tinkerbell of fairies, Captain Hook of pirates. Perfect princes and princesses are modeled on those in *Snow White*, *Cinderella*, and *Sleeping Beauty*. Our fears of a mother's death are partially molded by *Bambi*.

The Disney Mythology

The mythology disseminated by the Disney organization shows many similarities with myths of other cultures, and can be analyzed in the same anthropological, particularly structuralist, terms. In applying the structural method to Native American (that is, North and South American Indian) myths, Lévi-Strauss shows that binary oppositions are often resolved by mediating figures—entities that somehow link opposites. Consider an opposition, of nature to culture, that is a concern of humans everywhere. Zoologists and anthropologists know that many differences between humans and other animals are differences of degree rather than differences of kind, but religions and myths, for hundreds of thousands of years and throughout the world, have been concerned with demonstrating the opposite—that humans stand apart from nature, that people are unique. The opposition between the human—the cultural—and nature may be modeled in myth or religion on any one of the major attributes of culture: speech ("In the beginning was the word"), technology (Prometheus stole fire from the gods), thought (the soul), or knowledge (the fall of Adam and Eve). Human knowledge of good and evil is opposed to the innocence of animals.

In the myths that Lévi-Strauss has analyzed, oppositions are often resolved mythically by mediating figures. Animals, for example, are given human abilities, thus bridging the opposition between culture and nature. In Genesis a cultural, bipedal, talking, lying snake brings culture and nature closer together. Adam and Eve are, in the beginning, innocent parts of nature, yet they are unique because of their creation in God's image. The snake encourages original sin, which keeps humans unique, but in a far less exalted way. The punishment for eating forbidden fruit is physical labor; people are pitted in a struggle with nature. That humans are a part of nature, while at the same time different from other natural beings, is explained by the serpent-mediator's role in the fall. Note that the fall of humanity is also paralleled in the fall of the serpent—from culture-bearing creature to belly-crawling animal.

In Lévi-Strauss's terms, the principal feature or function of myth may be to mediate an opposition, to resolve a logical contradiction. Mediating figures and events may resolve such oppositions or contradictions as culture versus nature by showing that just as mythical animals can have human abilities and thus be cultural, people, while different from nature, are also part of nature. They are like animals in many ways, dependent on natural resources and participants in natural systems.

Interacting with larger trends of American life, Disney creations bridge the opposition between culture and nature. Disney confers a host of human attributes on his animated (*anima*, Latin "soul") nonhuman characters—talking, laughing, tricking, bumbling, lying, singing, friendship, family life. In most of his movies, in fact, the animals—and witches, dwarfs, fairies, pirates, and other not-quite-human characters—deny the opposition of culture and nature by having more human qualities than the stereotypically perfect heroes or heroines. Such "supporting characters" are remembered best because they are used not just to personify, but usually also to exaggerate, basic human imperfections. In *Snow White* a disguised and distorted witch-stepmother is evil incarnate. The dwarfs' names—Bashful, Sleepy, and the rest—read like a sanitized, child's version of the Seven Deadly Sins. *Lady and the Tramp* is a story about dogs who exaggerate American sex-role stereotypes of the 1950s.

In *Cinderella* the nature-culture opposition is inverted; mice—natural (undomesticated) animals that are ordinarily considered pests and inimical to humans—are endowed with speech and other cultural attributes and become Cinderella's loyal friends. The cat, ordinarily a part of culture (domesticated), becomes a dark creature of evil (named Lucifer) who almost blocks Cinderella's own transformation from domestic servant into princess. The transformation of Cinderella, a woman, is aided by *man's* best friend—a dog.

The reversal or inversion of the normal oppositions, that is, cat-culture-good versus mouse-nature-evil, demonstrates how, through Disney characterization, the opposition of culture and nature is overcome. Similarly, just as natural animals in Disney films are depicted as cultural creatures, people are often represented as closer to nature than we normally are. In several Disney films human actors are used to portray close relationships between children and such undomesticated animals as raccoons, foxes, bears, or wolves. His choice of Kipling's *The Jungle Book* as subject matter for a cartoon feature illustrates the second means of dealing with the nature-culture opposition.

A Pilgrimage to Walt Disney World

With Disney as creator and mythmaker for so many Americans, his shrines could hardly fail. In foreign cultures, many religious systems focus on sacred sites. Infertile Betsileo women in the highlands of central Madagascar seek fecundity by spilling the blood of a sacrificial cock before particular phallic stones (Kottak 1980). Native Australian totems are associated with holy sites where, in mythology, they first emerged from the ground. Sacred groves are shared reference points that provide symbolic unity for otherwise dispersed clans among the Jie of Uganda (Gulliver 1974). A visit to Mecca is an obligation of Islam. Miraculous cures are sought at shrines like Lourdes and Fátima associated with the mythology of Roman Catholicism. In the arid *sertão* of northeastern Brazil, more than twenty thousand pilgrims journey each August 6 to fulfill their vows to a wooden statue in a cave—Bom Jesus da Lapa. During the Bicentennial year, millions of Americans flocked to Philadelphia's Independence Hall, Washington's Lincoln and Jefferson Memorials and Washington Monument, to worship national ancestors, remember their achievements, and reaffirm national solidarity. Thousands of Americans waited in line in front of the National Archives to view a historical artifact—the Declaration of Independence—and to sign their names on a register to be put in a time capsule that will be opened in the twenty-first century. And virtually every day of every year, thousands of America families travel long distances, and often invest significant amounts of time and money, to experience Disneyland and Walt Disney World.

A conversation with anthropologist Alexander Moore first prompted me to think of Walt Disney World as analogous in some respects to the pilgrimage centers of the world's major religions, and the behavior, demeanor, and experiences of the millions of Americans who have visited or will visit it as comparable to the behavior of religious pilgrims. On the basis of his fieldwork at Disney World, Moore pointed out, for example, a striking similarity between that center and Mecca: division of the shrine into two distinct domains—an inner, sacred center and an outer, more secular domain. In Walt Disney World, appropriately enough, the inner, sacred area is known as "The Magic Kingdom."

Non-Disney motels, restaurants, and camping grounds dot the approach to Disney World; they become increasingly concentrated as one nears the park. As you enter the park, a few miles along World Drive, a sign on the right side of the road instructs you to turn to a specified AM radio station. A recording played continuously throughout the day gives information about where and how to park and how to proceed on the journey to the Magic Kingdom. It also promotes new Magic Kingdom activities and special attractions such as "America on Parade" and "Senior American Days." Travelers enter the mammoth parking lot by driving through a structure like a turnpike toll booth. As they pay the parking fee, they receive a brochure describing Disney attractions located both inside and outside the central area. (Camping grounds, lakes, islands, and an "international shopping village" are located in the park's outlying areas.) Sections of the parking lot bear totemlike designations: Minnie, Goofy, Pluto, and Chip 'n' Dale, each with several numbered rows. Uniformed attendants direct drivers to parking places, making sure that cars park within the marked spaces and that every space is filled in order. As visitors emerge from their cars, they are directed to open-air, trainlike buses called "trams." Lest they forget where their automobiles are parked, they are told, as they board the tram, to remember Minnie, Pluto, or whichever mythological figure has become temporary guardian of their car. Many travelers spend the first minute of their tram ride ritually reciting "Minnie, Minnie, Minnie, Minnie," memorizing their automobile's location. Leaving the tram, visitors hurry to booths where they purchase ticket booklets granting entrance to the Magic Kingdom and its attractions ("adventures"). They then pass through turnstiles behind the ticket sales booths and prepare to be transported, by "express" monorail or ferryboat, to the Magic Kingdom itself.

Because the approach to the central area occurs in gradual, definite stages, the division of Walt Disney World into outer, secular space and inner, sacred space is not clear-cut. This gradual entry into the center of Walt Disney World can be compared to Judaism's conception of the zonation of sacred space radiating out from Jerusalem. In orthodox Judaism, Jerusalem is conceived as the center (navel) of the world. In the center of Jerusalem is the Temple, in the center of which is the altar, in the center of which is the ark, and in front of which is the foundation stone through which humans are linked to heaven. (An interesting discussion of the relationship between this religious world view and the Sunday morning bagel ritual of many Jewish Americans can be found in Regelson 1976.) Sanctity is concentrated at the center, and the zones closing concentrically inward become gradually, rather than abruptly, more sacred. In the central Disney World area, even after one has passed parking lot and turnstiles, a still secular zone, where central hotels, beaches, and boating areas are located, intervenes before the Magic Kingdom. Visitors can take a "local" monorail to a hotel, but most choose the express monorail to the Magic Kingdom, a futuristic mode of transportation, rather than the alternative, the ferryboat.

Aboard the express monorail, which accomplishes for most visitors the transition between outer, more secular areas and the Magic Kingdom, similarities between Walt Disney World pilgrims and participants in passage rites become especially obvious. Rites of passage, remember, involve transitions in space as well as in age or social status. Disney pilgrims who ride the express monorail exhibit, as one might expect of a transition from secular to sacred space (a magic kingdom), many of the attributes that Turner (1974) associates with liminal states. Like liminal periods in other passage rites, during the transition aboard the monorail all prohibitions that apply anywhere else in Disney World are intensified. In the secular areas and in the Magic Kingdom itself, people are allowed to smoke and eat, and in the secular areas they can consume alcohol, have sexual intercourse (in hotel rooms), and go shoeless—but all these things are tabooed aboard the monorail. Like liminal travelers through other passage rites, Disney pilgrims temporarily relinquish control over their own destinies. Herded like cattle through gates leading to sections of the monorail, passengers enter a place apart from ordinary space, a time out of time, in which social distinctions disappear, in which everyone is reduced to a common level. As the monorail departs, a disembodied voice prepares the pilgrims for what is to come, enculturating them in the lore and standards of Walt Disney World.

Typical, too, according to Turner (1974), of liminal periods are reversals of the sexual aspects of everyday secular life and, at the end of liminality, the appearance of symbols of the passenger's ritual rebirth. Extraterrestrial anthropologists studying behavior on the monorail might note that although sexual abstinence is prescribed, major sexual and rebirth symbolism is an aspect of the ride. As the express monorail speeds *through* the Contemporary Resort Hotel, travelers facing forward observe an enormous tiled mural that covers an entire wall of the hotel. Just before the monorail reaches the hotel, but much more clearly after it emerges, travelers espy Walt Disney World's primary symbol, Cinderella's castle. In Freudian terms, this juxtaposition of castle and penetration might well be seen as a symbolic representation of sexual intercourse; and the emergence from the mural into full view of the Magic Kingdom might be viewed as a simulation of rebirth. The monorail, one might note, is simultaneously male and female. It is phallic-shaped, yet at the same time is the womb in which passengers are carried from a secular state to rebirth in the sacred domain.

Many readers may find this symbolic analysis farfetched, but it is included to illustrate, using American natives as examples, techniques that anthropologists use to describe symbols associated with religious rituals in our own and other societies. As in the structural analysis of rock and football, such analyses of symbols permit a variety of alternative interpretations, and it is difficult to judge which is best. However, even if this analysis of monorail symbolism is rejected, other parts of the analysis confirm similarities between visits to Walt Disney World, on the one hand, and "religious" pilgrimages and passage rites, on the other.

Within the Magic Kingdom

Once the monorail pulls into the Magic Kingdom station, the transition is complete. Passengers emerge on their own; direction by attendants, so prominent at the other end of the line, is conspicuously absent. Walking down a ramp, travelers pass through another turnstile; a transit building where lockers, phones, rest rooms, strollers, and wheelchairs are available; and a circular open area. They find themselves in the Magic Kingdom, walking down "Main Street, U.S.A."

The Magic Kingdom itself invites comparison with shrines and ostensibly religious rites. A. R. Radcliffe-Brown (1952) and several other anthropologists have argued that the major social function of religious rites is to reaffirm and thus to maintain social solidarity among participants, members of the congregation. Turner (1974) suggests that certain rituals among the Ndembu of Zambia serve a *mnemonic* function (they make people remember). Women's belief that they can be afflicted and made ill by spirits of their deceased kinswomen leads them to take part in rites that reaffirm the values of their kinship system by reminding women of their ancestors. Ancestor worship in any society serves a similar mnemonic purpose.

Similar observations can be made about Walt Disney World. Frontierland, Liberty Square, Main Street, U.S.A., Tomorrowland, and Fantasyland—the major sections of the Magic Kingdom—not only remind pilgrims of the important historical values on which America is purported to be founded, but also juxtapose and link together, through the creative vision of Walt Disney, the past, present, and future; childhood and adulthood; the real and the unreal. Many of the adventures, or rides, particularly Space Mountain—an innovative but frightening roller coaster—can be compared to anxiety-producing rites. This anxiety is ultimately dispelled when pilgrims realize that they have survived simulated speeds of ninety miles an hour.

Detaching oneself from American culture, one might ask how a Betsileo visitor from Madagascar would view Disney World adventures, particularly those based on fantasy. Among the Betsileo, and in many other nonindustrial societies, witches are actual people—parts of reality rather than of fantasy. Peasants in Brazil and elsewhere believe in witches, werewolves, and evil creatures of the night. A Betsileo would probably find it difficult to understand why Americans voluntarily take rides designed to produce uncertainty and fright.

Pilgrims agree implicitly, during their visit to Disney World, to constitute a temporary community, to spend a few hours or days observing the same rules, sharing experiences, and behaving alike. They share a common social status as pilgrims, waiting for hours in line, presenting tickets to attendants.

Yet the structure and attractions of the Magic Kingdom also relate to higher levels of sanctity, representing, recalling, and reaffirming not only Walt Disney's creative acts, but values of American society at large. In the Hall of Presidents, considered one of the Magic Kingdom's major attractions, people silently and reverently view a display of our national ancestors—moving, talking, lifelike dummies. Like Ndembu rites, aspects of the Magic Kingdom make us remember. Not only are we made to recall presidents and American history, but such familiar characters in children's literature as Tom Sawyer are brought to mind. And, of course, we meet the cartoon characters who, in the person of costumed humans, walk around the Magic Kingdom, posing for photographs with children. The juxtaposition of past, present, future, and fantastic represents eternity and argues that our nation, our people, our technological expertise, and our beliefs, our myths, and our values will endure. Even dress codes for Disney World employees reaffirm the stereotype of the clean-cut American. The whole ambience of Disney World reiterates the value of coupling industrial technology with such character traits as clean living, enthusiasm, imagination, creativity, and vision. Disney propaganda uses Walt Disney World itself to illustrate what American creativity joined with technical know-how can accomplish. Students in American history are told how our ancestors carved a new land out of wilderness; similarly, Walt Disney is presented as a person who created cosmos out of chaos, a structured world out of the undeveloped chaos of Florida's central interior.

A few other links between Walt Disney World and religious and quasi-religious symbols and shrines should be examined. Disney World's most potent symbol is Cinderella's castle, complete with a moat where pilgrims throw coins and make wishes. On my first pilgrimage to Walt Disney World I was surprised to discover that the castle has a largely symbolic function—as a trademark or logo for Walt Disney World—and little utilitarian value. A few shops on the ground floor are open to the public, but the rest of the castle is off limits. In thinking about the function of Cinderella's castle, I recall a lecture given at the University of Michigan in 1976 by British anthropologist Sir Edmund Leach. In describing the ritual surrounding his dubbing as knight, Leach noted that Queen Elizabeth stood in front of the British throne and did not, in accordance with our stereotype of monarchs, sit on it. Leach concluded that the primary value of the British throne is to represent, to make concrete, something enduring but abstract—the British sovereign's right to rule. Similarly, Cinderella's castle offers concrete testimony to the eternal aspects of Walt Disney creations.

A Pilgrimage to a "Religious" Shrine

A brief comparison of Walt Disney World with Bom Jesus da Lapa, a shrine center in the arid backlands of Bahia state, Brazil, will point out further similarities between the former and "religious" pilgrimage sites. Normally a town of some eight thousand people, Bom Jesus da Lapa is isolated like Walt Disney World in the interior of its state but experiences an annual influx of more than twenty thousand pilgrims on August 6. The patron saint, Bom ("Good") Jesus, is represented by a wooden statue that stands atop an altar in a cave. Like Cinderella's castle, a well-known landmark—a gray limestone outcrop pitted with caves—identifies Bom Jesus to pilgrims. Gross (1971) compares this symbol to a medieval fantasy.

Gross found that most pilgrims had come to Bom Jesus da Lapa to fulfill vows, usually concerned with health. They pray to Bom Jesus and promise to make the pilgrimage to his shrine if their request is granted. Prayer requests are varied. Bom Jesus may be asked to help cure a specific malady, to guarantee a safe journey, to enable lovers to stop quarreling. To further fulfill their vows, pilgrims make offerings at the altar. If the prayer concerned a successful marriage, a photograph of the happy couple may be offered. People who have prayed for a broken leg to heal may leave X-rays or casts of the leg at the altar of Bom Jesus.

The kinds of reasons people give for making pilgrimages vary from shrine to shrine. Most people go to Bom Jesus da Lapa to fulfill vows made in prayers. A visit to Mecca is an obligation of Islam. Miraculous cures are sought, and reported, at Lourdes, just as during performances by itinerant faith healers in the contemporary United States. Interviews with visitors to Walt Disney World would probably reveal a similar variety of motives for making the trip. "Pleasing the children" would probably be a frequent reason. Also, parents offer a trip to Disney World as a major reward for children's behavior and achievements or, perhaps, as an incentive to help them recover from long illness. Most Americans probably visit Walt Disney World for amusement, recreation, and vacation. In this sense, they differ from pilgrims to religious shrines. They do not appear to believe that a visit to Disney World has curative properties, although they may feel that vacations are good for their health. Nonetheless, local television news programs have presented human interest stories about communities pooling their resources to help terminally ill children and their families visit Disney World or Disneyland. Thus, although a visit to Walt Disney World is not conceived by Americans as curative, it is regarded as being an appropriate last wish.

Even when people undertake "religious" pilgrimages, however, their motives may not be exclusively or even primarily "religious," as Gross's account of the Bom Jesus da Lapa pilgrimage illustrates. Because there are so many pilgrims to Bom Jesus, most have no chance to accomplish the ostensible purpose of their journey, to worship the wooden statue of Bom Jesus da Lapa. They are rapidly herded by chapel officials past the altar, just as Disney visitors are corralled into tram and monorail. Many pilgrims are forced to move on to make way for others before they have a chance to kneel. Furthermore, Bom Jesus and Disney World have similar commercial and recreational aspects. A variety of souvenirs, not limited to church-related icons, are sold to Bom Jesus pilgrims, as in Disney World. In fact, the Bom Jesus pilgrim spends little time in religious contemplation. Several kinds of entertainment come to Bom Jesus along with the pilgrims: traveling circuses, trained boa constrictors, vaudeville acts, roulette wheels and other gambling devices, and singing troubadors. During the height of the pilgrimage, Bom Jesus sports more than a dozen brothels. In fact, Americans might find Walt Disney World, with its celebration of the values of clean living, more ascetic and religious than the Bom Jesus "religious" shrine. It is not unlikely that similar nonreligious activities and a similar representation of other-than-religious motives characterize popular shrines and pilgrims elsewhere.

Many Americans believe that recreation and religion are distinct and incompatible domains. On the basis of my own fieldwork in Brazil and Madagascar, and from reading about other societies, I believe this separation to be both ethnocentric and false. Betsileo tomb-centered ceremonials are times when the living and the dead are joyously reunited; when people get drunk, sing and dance, gorge themselves, and have sexual license. Perhaps the gray, sober, and ascetic aspects of many religious events in the contemporary United States, in taking the fun out of religion, force us unwittingly to find our religion in fun—to seek in such apparently secular contexts as Disney amusement parks, rock concerts, and football games what other people find in their rites, beliefs, and ceremonies.

This account of Disney World exemplifies an etic analysis. Walt Disney has not been viewed as a merely commercial figure nor his amusement parks as simply recreational domains. Rather, through a measure of anthropological detachment, a deeper level of attachment between Americans, Walt Disney, and his creations has been described. The implication is not that Americans' relationship to Disney constitutes a religion. But through observation of behavior and beliefs, and through comparison with pilgrims in other parts of the world, similarities between passage rites, religious pilgrimages, and rituals, on the one hand, and our relationship to Disney, on the other, have been noted. Disney, his amusement parks, and his creations constitute a powerful complex of enculturative forces in contemporary American society. Still another example of common American ritual can be found in an even more unlikely commercial setting, the fast-food restaurant chain.

McDonald's as Myth, Symbol, and Ritual

Each day, on the average, a new McDonald's restaurant opens somewhere in the world. The number of McDonald's outlets today surpasses the total number of fast-food restaurants in the United States in 1945. In barely twenty-five years McDonald's has grown from a single hamburger stand in San Bernardino, California, into today's international web of outlets—more than four thousand of them—located throughout the United States and in such foreign countries as Japan, Hong Kong, Mexico, France, Sweden, England, Germany, and Australia. In 1972, annual sales of $1.3 billion propelled McDonald's past Kentucky Fried Chicken as the world's most successful fast-food chain, a position it has held ever since. Today's sales exceed $3 billion a year.

Have factors less obvious to American natives than relatively low cost, fast service, cleanliness, and taste contributed to the success of McDonald's? Could it be that natives—in consuming the products and propaganda of McDonald's—are not just eating or watching television, but are experiencing something comparable in certain respects to participation in religious rituals? To answer this question we must briefly consider the nature of ritual.

Anthropologist Roy Rappaport (1974) mentions several features that distinguish rituals from other behavior. Rituals, first, are formal—stylized, repetitive, and stereotyped. They are performed in special (sacred) places and occur at set times. Rituals include liturgical orders—set sequences of words and actions laid down by someone other than the current performers. All these features link rituals to plays, but there are important differences. Plays have audiences rather than participants; actors are only *portraying* something, whereas ritual performers—who make up congregations—are *in earnest*. Rituals also convey information about participants and their cultural traditions. Performed year after year, generation after generation, rituals translate into observable action enduring messages, values, and sentiments. Rituals are social acts. Although inevitably some participants are more strongly committed to the beliefs on which the rituals are founded than others, just by taking part in a joint public act, the performers signal that they accept an order that transcends their status as mere individuals.

In view of some anthropologists, including Rappaport himself, such secular institutions as McDonald's are not at all comparable to rituals. Rituals, they argue, involve special emotions, nonutilitarian intentions, and supernatural entities, which are not characteristic of

Americans' participation in McDonald's. But other anthropologists define ritual more broadly. Writing about football, Arens (1976) points out that behavior can simultaneously have sacred and secular aspects. Thus, on one level, football can be seen as simply a sport, while on another it can be seen as a public ritual. On one level McDonald's is a mundane, secular institution—just a place to eat—but on another it assumes some of the attributes of a sacred place. And, in the context of comparative religion, why should this be surprising? The French sociologist Emile Durkheim (1954, 1st ed. 1915) pointed out long ago that almost everything, from the sublime to the ridiculous, has in some societies been treated as sacred. The distinction between sacred and profane does not depend on the intrinsic qualities of the sacred symbol. In Australian aboriginal totemism, for example, Durkheim found that sacred entities were generally such humble and nonimposing creatures as ducks, frogs, rabbits, and grubs—beings whose inherent qualities could hardly have been the origin of the religious sentiment they inspired. If frogs and grubs can be elevated to a sacred level, why not McDonald's?

Behavioral Uniformity at McDonald's

For several years, like many other Americans, I have frequently had lunch and occasionally had "dinner" at McDonald's. Recently, I have begun to observe carefully certain rituallike aspects of Americans' behavior at these fast-food restaurants. Tell your fellow Americans that going to McDonald's is similar in some ways to going to church or temple and their emic bias as natives will reveal itself in laughter, scorn, denial, or questions about your sanity. Just as football is a game and Walt Disney World an amusement park, McDonald's, for natives, is just a place to eat. However, a comparison of what goes on in McDonald's outlets in this country will reveal a very high degree of formal, uniform behavior from store to store on the part of staff and customers alike. Particularly interesting is that this invariance in act and utterance has developed in the absence of any theological doctrine. McDonald's ritual aspect, rather, is founded on twentieth-century technology, particularly on automobiles, television, work away from home, and the one-hour lunch break. It is striking, nevertheless, that one commercial organization should be so much more successful than other businesses, the United States Army, and even many religious institutions in producing behavioral invariance. Factors other than low cost, fast service, and the taste of the food—all of which are approximated by other chains—have contributed to our acceptance of McDonald's and our adherence to its rules.

Remarkably, when Americans travel abroad, even in countries noted for distinctive cuisines, many visit the local McDonald's outlet. It seems that the same factors that lead us to return to McDonald's again and again in our own country are responsible. Because Americans are thoroughly familiar with how to behave, what to expect, what they will eat, and more or less what they will pay at McDonald's, in its outlets overseas they have a kind of home away from home. In Paris, whose natives are not especially renowned for making tourists, particularly Americans, feel at home, McDonald's offers sanctuary. It is, after all, a uniquely American institution, where only natives, programmed by years of prior experience, can feel completely at home. Americans, if they wish, can temporarily reverse roles with their hosts; if American tourists can't be expected to act like the French, neither can the French be expected to act in a culturally appropriate manner at McDonald's.

This devotion to McDonald's rests in part on uniformities associated with almost all its outlets, at least in the United States: food, setting, architecture, ambience, acts, and utterances. For example, the McDonald's symbol, its golden arches, is an almost universal landmark, just as familiar to Americans as Mickey Mouse, Miss Piggy, and the Beatles. Few McDonald's outlets in this country lack golden arches as outside markers. One of them, where a significant portion of my fieldwork has been done, is in Ann Arbor, a block and a half from the University of Michigan. However, the absence of this symbol does not reduce the outlet's impact as a quasi-ritual setting. Although the restaurant is a contemporary brick structure, it has stained glass windows, with golden arches as their central theme. Sunlight floods in through a skylight that recalls the clerestory of a church. The golden arches are uniform symbols that are universally recognized by, and have special significance for, participants. Thus analogies between religious symbols and the arches are not, I think, farfetched. And in Ann Arbor, where the golden arches are absent, McDonald's stained glass windows and clerestory certainly conjure up religious connotations among natives I have interviewed.

Americans enter a McDonald's restaurant to perform an ordinary, secular act—to eat, usually lunch. Yet the surroundings there tell us that we are somehow apart from the variety, disorder, and variability of the world outside. We know what we are going to see, what we are going to say, what will be said to us, what we will eat, how it will taste, and how much it will cost. Behind the counter at McDonald's, agents are differentiated into three categories: male help, female help, and managers.

Members of each group wear similar attire. Permissible utterances by customer and worker alike are liturgically specified above the counter. Throughout the United States, with only minor variation, the menu is located in the same place, contains the same items, and has the same prices. The food, again with only minor regional variation, will also be prepared according to plan, and will vary little in taste. Obviously customers are limited in what they can choose; less obviously, they are also limited in what they can say. Each item has its appropriate McDonald's designation: "large fry," "quarter pounder with cheese." The neophyte customer who innocently asks, "What kind of hamburgers do you have?" or "What's a Big Mac?" is out of place.

Other ritualized phrases are uttered by the person behind the counter. If a man asks for a quarter pounder, the automatic response is, "Will that be with cheese, sir?" After the customer has completed his order, if no potatoes have been requested, the agent ritually incants, "Will there be any fries today, sir?" Once food is presented and picked up, the agent conventionally says, "Have a nice day." Nonverbal behavior of McDonald's agents is programmed as well. Before opening the spigot of the drink machine, workers fill paper cups with ice exactly to the bottom of the golden arches that decorate them. As customers request food, agents look back to see if the desired sandwich item is available. If not, they tell you, "That'll be a few minutes" and prepare your drink, after which a proper agent will take the order of the next customer in line. McDonald's lore of appropriate verbal and nonverbal behavior is even taught at a "seminary," called Hamburger University, located in Illinois. Managers who attend pass the program they have learned on to the people who work in their restaurants.

It is not simply the formality and regularity of behavior at McDonald's but its total ambience that invites comparison with sacred places. Like the Disney organization, McDonald's image-makers stress clean living and draw on an order of values—"traditional American values"—that transcends McDonald's itself. Agents submit to dress codes and are uniformly and cleanly attired. Their hair length, height, and complexions are scrutinized by management. McDonald's kitchens, grills, and counters sparkle. Styrofoam food containers that promise to haunt the world for eons are used only once. Understandably, the chain's contributions to worldwide product pollution (along with labor practices that have been questioned) evoke considerable hostility. In 1975 the Ann Arbor campus McDonald's was the scene of a ritual rebellion—desecration by the Radical Vegetarian League, who held a "puke-in." Standing on the second-story balcony just below the clerestory, more than a dozen vegetarians gorged themselves on mustard and water and vomited down on the customer waiting area. McDonald's, defiled, lost many customers that day.

The formality and relative invariance of behavior in a ritually and symbolically demarcated setting thus suggests some analogies between McDonald's and the sacred. Furthermore, like performance of a ritual, participation in McDonald's occurs at specified times fixed by clock, calendar, or specified circumstances. In American culture our daily food consumption is supposed to occur as three meals: breakfast, lunch, and dinner, of which dinner is stressed as the major meal. Americans who have traveled abroad are aware that cultures differ in which meal they emphasize. In many parts of the world, the midday meal is primary. Americans are away from home at lunchtime because of their jobs, and are usually allowed only an hour for lunch; they view dinner as the main meal, and lunch as a lighter meal symbolized by the sandwich. McDonald's provides relatively hot and fresh sandwiches and a variety of subsidiary fare that many American palates can tolerate.

The ritual of eating at McDonald's is confined to ordinary, everyday life. Eating at McDonald's and religious feasts are in complementary distribution in American life; that is, when one occurs the other does not. Most Americans would consider it inappropriate to eat at a fast-food restaurant on Christmas, Thanksgiving, Easter, Passover, or other religious and quasi-religious feast days. Such feast days are often holidays from work and are regarded in our culture as family days, occasions when relatives and close friends get together. However, although Americans neglect McDonald's on holidays, television reminds us that McDonald's still endures, that it will welcome us back once our holiday is over. The television presence of McDonald's is particularly obvious, in fact, on such occasions—whether through a float in the Macy's Thanksgiving Day parade or through sponsorship of special programs, particularly of "family entertainment."

McDonald's Advertising

Although such chains as Burger King, Burger Chef, Wendy's, and Arby's compete with McDonald's for the fast-food business, none rivals McDonald's success. The explanation probably lies in the particularly skillful ways in which McDonald's advertising plays up the features just discussed. Its commercials are varied to appeal to different audiences. On Saturday morning television, with its steady stream of cartoons and other children's features, McDonald's is a ubiquitous sponsor. The McDonald's commercials for children's shows usually differ from the ones that adults see in the evening and during football games. Children are introduced to, and reminded of, McDonald's through several fantasy characters, headed by the clown Ronald McDonald. These commercials often stress the enduring aspects of McDonald's. For example, Ronald has a time machine that enables him to introduce hamburgers to the remote past and the distant future. Anyone who noticed the McDonald's restaurant in the Woody Allen film *Sleeper*, which takes place 200 years hence, will be aware that the message of McDonald's as eternal has gotten across. As in Disney films, but on a limited and less dramatic level, representations of conflict between good (Ronald) and evil (Hamburglar) are gently portrayed. Children can meet the "McDonaldland" characters again at McDonald's outlets. Their pictures appear on McDonaldland cookie boxes, and, from time to time, on glasses and plastic cups that are given away with the purchase of a soft drink. Even more obvious are children's chances to actually meet Ronald McDonald. Actors portraying Ronald scatter visits, usually on Saturdays, among McDonald's outlets throughout the country. One can even rent a Ronald for a birthday party.

McDonald's adult advertising has a different but equally effective theme. In 1976 and 1977 breakfast at McDonald's was promoted by a fresh-faced, sincere, happy, clean-cut young woman, inviting the viewer to try a new meal in a familiar setting. In other commercials, healthy, clean-living Americans gambol on ski slopes or in mountain pastures. The single theme, however, that runs through all the adult commercials is personalism. McDonald's, the commercials drone on, is something other than a fast-food restaurant. It is a warm, friendly place where you will be graciously welcomed and will feel at home, where your children won't get into trouble. Other chains have copied McDonald's incessant emphasis on the word *you*: "You deserve a break today"; "You, you're the one"; "We do it all for you." McDonald's commercials tell you that you are not simply an anonymous face in an amorphous crowd. You can find respite from a hectic and impersonal society, the break you deserve. Your individuality and dignity will be respected in McDonald's. McDonald's backpacks take up the theme. "Me, I'm the One," they say. You become an individual—one with yourself—you are told, by becoming one with McDonald's. Verbally and nonverbally you demonstrate your participation in the subculture of McDonald's. You load your books in a McDonald's backpack, you impress your friends by your ability to recite "Two all beef patties, special sauce, lettuce, cheese, pickles, onions on a sesame seed bun."

McDonald's advertising tries to deemphasize the fact that the chain is, after all, a commercial organization. In the jingle, you heard, "You, you're the one; we're fixin' breakfast for ya"—not, "You, you're the one; we're makin' millions off ya." McDonald's is presented as much more of an American institution than the apple pie it serves. Commercials make it seem like a charitable organization by stressing its program of community good works. During the Bicentennial year, commercials reported that McDonald's was giving 1,776 trees to every state in the union. How big the trees were, or what the states did with them, was never specified. Brochures at outlets echo the television message that, through McDonald's, you can sponsor a carnival to aid victims of muscular dystrophy. Again in 1976, McDonald's sponsored a radio series documenting contributions to American history of Afro-Americans. Such "good, clean" family television entertainment as the film *The Sound of Music* was brought to you by McDonald's, complete with a prefatory sermonlike address by its head, Ray Kroc. On this occasion, special commercials united Ronald McDonald (shown picking up after litterbugs) with the adult commercial themes. McDonald's commercials tell us that it supports and works to maintain the values of American family life. They have even suggested a means of strengthening what most Americans conceive to be the weakest link in the nuclear family: father-child. "Take a father to lunch," kids are told; love your father at McDonald's.

As with the Disney organization, the argument here is certainly *not* that McDonald's has become a religion. Rather, it is merely being suggested that specific ways in which Americans participate in McDonald's bear analogies to religious systems involving myth, symbol, and ritual. Just as in rituals, participation in McDonald's involves temporary subordination of individual differences in a social and cultural collectivity. By eating at McDonald's we communicate information about our current physical state (hunger). Even more important, by going to McDonald's, or by wearing McDonald's backpacks, we convey information about ourselves to others. In a land of tremendous ethnic, social, economic, and religious diversity we demonstrate that we share something with millions of other Americans. Furthermore, as in ritual performances, participation in McDonald's is linked to a cultural system that transcends McDonald's itself. By eating at McDonald's we say something about ourselves as Americans, about our acceptance of values and ways of living that belong to a social collectivity. By returning to

McDonald's over and over again, we affirm that certain values and lifestyles, developed through the collective experiences of Americans before us, will continue.

Anthropology and American "Pop" Culture

Just as etic analysis of religious ceremonials in nonindustrial societies may demonstrate that they serve ecological and economic functions that are unrecognized by natives, etic analysis of Americans' participation in, and beliefs and feelings about, football, rock music, Disney creations, and McDonald's fast-food outlets reveals unsuspected analogies with religious rituals, myths, and symbols.

The growing anthropological interest in American society has prompted studies of both variation and uniformity. For example, aspects of social and cultural variation have been linked to such factors as socioeconomic class, poverty, ethnicity, and rural versus urban residence. In this essay, however, uniformity has been the focus. The stress on experiences and enculturative forces common to most Americans, particularly the young, has emphasized several points.

Anthropology is not simply the study of nonindustrial populations. Many techniques that anthropologists have applied to other cultures—including structural, symbolic, and etic analyses—can be used just as easily to interpret American culture. In studies of their own cultures, native anthropologists can contribute uniquely by coupling detachment and objectivity with their own experience and understanding as natives.

This essay has also demonstrated that structural and symbolic analyses of aspects of culture share as much with the humanities as with the sciences. These approaches seek primarily to *discover*, *rethink*, and *illuminate* otherwise hidden dimensions or deeper meanings of phenomena rather than to explain them. Structural and symbolic analyses of any culture are, therefore, difficult to confirm or to falsify. They can be evaluated emically: Do natives accept them or prefer them to alternative interpretations? Do they enable natives to make more sense of familiar phenomena? They can also be evaluated etically: Do they fit within a comparative framework provided by data and analyses from other societies? However, many anthropologists remain skeptical about the value of such impressionistic approaches, preferring research strategies that may be evaluated statistically and confirmed or denied by researchers independently examining the same data.

The examples considered in this essay are new and uniquely

American shared cultural forms that have appeared and spread rapidly during the twentieth century because of major changes in the material conditions of American life—particularly work organization, communication, and transportation. Late-twentieth-century Americans deem at least one automobile a necessity, and televisions now outnumber toilets in American households. Through the mass media, such institutions as football, rock music, Disney creations, and McDonald's have become powerful elements of American national culture, providing a framework of common expectations, experiences, and behavior that overrides region, class, formal religious affiliation, political sentiments, gender, ethnic group, and place of residence. Although some of us may not like these changes, we certainly can't deny their significance.

The rise of fast-food restaurants, Disney, rock, football, and similar institutions is related not just to the mass media but also to the decreasing participation in traditional organized religion, and the weakening of ties based on kinship, marriage, and community within industrial society. Neither formal religion, strong centralized government, nor kinship unite most Americans. Cars, computers, movies, television, stereos, and their by-products do, and in this Americans provide a uniquely exotic example in the realm of cultural diversity.

These aspects of contemporary American culture are perhaps viewed as merely passing, or "pop," culture by certain segments of American society, but they are highly significant features of our culture, shared by millions of Americans. As such, they certainly deserve, and are receiving, the attention of anthropologists and other scholars. Such studies are fulfilling the traditional promise offered in textbooks that by studying anthropology we can learn more about ourselves. Americans can view themselves not just as members of a varied and complex nation, but also as a population united by distinctive shared symbols, customs, and experiences. Although American culture resembles others in certain respects, like any other cultural system, it also makes its own unique and original contributions to the realm of cultural diversity. And that, after all, is the subject matter of anthropology.

Professor McKinley analyzes how television newscasts reflect a view of the world (cosmology) characteristic of American culture. Adopting a structuralist stance, he argues that the familiar sequencing of news, sports, and weather mirrors the cultural contrast between culture (news) and nature (weather) with sports ("which display cultural plays on natural abilities") as mediator. McKinley also notes that the 6:00 P.M. newscast, the day's most important, mediates between the world of work (business time) and the world of play (leisure time). He contends, too, that American cosmology associates the East with culture (and standards, as in Eastern Standard Time) and the West with nature (Mountain and Pacific Times are names drawn from nature—mountains and ocean). Weather (nature) moves from west to east, whereas culture (true culture, not "play culture," which can originate in California) moves from east to west. McKinley's stimulating paper makes several other points about ritual, symbols, and structuring in "the news."

4 / Culture Meets Nature on the Six O'Clock News: An Example of American Cosmology

Robert McKinley

The word *news* means "a report of a recent event" (*Webster's New Collegiate Dictionary*). News is information about recent events, and it is information which is told. To think about news is to make important assumptions about time, about events, and about ways of telling things to people. And these assumptions have their roots in socially defined reality. They reflect the way a society or culture structures its view of the world.

Events do not locate themselves. They belong to the flux of experience, and if they are to be singled out for discussion, then certain structured notions must be relied upon in order to place events in context and give them meaning. Two general types of structure provide the contexts for interpreting newsworthy events: (1) the structure of society itself, and (2) the structure of the symbolic form, or news text, through which the news is told.

Structures of the first type are illustrated by the fact that if a president of the United States were to die in office this would be regarded as a more newsworthy event than, say, my own death should I be run over by a car on my way home from work. American social structure determines the difference between these two events. The second type of structure is illustrated by the fact that different news formats are used to report different kinds of events. A front page headline would be used to tell of a president's death, whereas my own untimely demise would receive only a mention on one of the back pages. That different typefaces and sections of newspapers exist is part of the structure of the printed text, not part of the structure of society as such.

Examples of these two kinds of structure can be multiplied to infinity. But the point has been made: news reporting offers a double illustration of the inability of events to define themselves. Any event in the news derives its meaning from two separate structural contexts: (1) its original social locus, and (2) the symbolic form in which it is finally reported.

This essay is concerned with the relations between these two types of structures. Specifically, it asks how general assumptions about the makeup of social reality get expressed in typical news formats and texts. The example taken up here is American television news broadcasts. Much can be asked of these texts. Why, for example, is it common for local TV news programs to present the events of the day in a standard sequence of first the news, second the sports, and third the weather? Is some important cultural paradigm contained within this pattern? Also, why, when women broadcasters are present, do they more often give the weather than either the news or the sports? Finally, is anything intended by departures from these patterns?

It must be kept in mind that these are questions regarding the extent to which very broad cultural assumptions enter into the meaning of TV news formats. They are not questions about editorial bias or the distortion of facts in reporting the news. Nor are they questions of the extent to which the news medium itself becomes news. That news texts can misrepresent the social relevance of events reported, or that they can become so autonomous as to create what are now called "media events" are significant problems and ones that have received much comment from critics of the media. However, these are problems of a different order from the one I am tackling. My concern is not with the accuracy or appropriateness of the way events are reported but with the fact that they cannot be reported at all unless a symbolic framework is employed. That framework must be a fairly stable one. And to be stable it must draw upon key assumptions that American culture holds regarding the more permanent structures of reality. Given that news formats are symbolic structures, and that they are presented at very regular times, they have a certain ritual quality about them. The difference between the cultural analysis I offer and other forms of commentary on the media is that I assume this ritual quality to be as important to the news text as is the news itself.

In short, I am not examining distortion in news reporting so much as I am examining how the nightly ritual of presenting the news at all reaffirms the reality status of certain ideas central to American cosmology. These ideas concern time and space and the opposition between nature and culture.

Date Line, Now

Cultural aspects of time are inescapable features of TV news formats. News shows are aired at times corresponding to morning, noon, evening, and night. One can think of this as breakfast, lunch, dinner, and bedtime. Just as dinner is the ritually most important meal of the day, so too the ritually most important of these news presentations is the evening news. In many local areas 6:00 P.M. is the starting time for such broadcasts. A reason why the six o'clock news is the most important is that it is the most immediate full tally of the day's events. The morning news is left over from the day before. The noon news is not complete enough to seem final, or authoritative, and the late news is merely a kind of recap before ending the day of things covered in greater depth at six o'clock. Between the six o'clock news and the late news (aired at either ten or eleven o'clock in different parts of the country) not much of official public importance is expected to occur because this is after business hours. So we see that the prominence of the six o'clock news is linked to the important opposition in American daily lives between work time and leisure time.

However, in connection with this opposition between work time and leisure time there is a more complex aspect of "news time" itself which accounts for the preeminence of the six o'clock news. This goes right back to the definition of news as "a report of a recent event." Relative to the full business day, the six o'clock news is the most recent of the four daily TV news broadcasts. News and time are linked propositions, just as are history and time. But news is in some ways a kind of reverse history. The value given to public knowledge of events in news reporting is the opposite of that given in history and mythology. In history that which happened first in time has priority. In myth this extends to time immemorial. But in news reporting it is only what has happened last that stimulates interest. The most exciting news is the latest news, and competitors in the news business are always out to scoop each other in knowing the latest developments in a story. When events reported are no longer recent, they are no longer news. The sense of time in the news is very rapid. The sense of knowledge is that of unending change. These qualities of the news are often symbolized in the pulsating electronic sounds or rhythmic music that is produced at the beginning of a news broadcast. Somehow beeps and flashes are appropriate to the idea that reports of recent events are reports of an unending change, unending change that is made to seem more immediate and up-to-date by these sounds.

Relative to the complete working day, it is the six o'clock news that is most recent, so it is presented as the fullest news report of the day. This fullness is shown by the back-to-back airing of a half-hour- or hour-long local news program with that of a national network broadcast. All the strictly news, sports, or weather segments of the morning, noon, and late-night news are either shorter than those of the six o'clock news or else they lack the nationwide programming of the latter. To this should be added the fact that the nationwide program is thought to be devoted to "world and national news," while the local programs are devoted to "local news." All this serves as a kind of collective representation of town, state, nation, and globe. But in this sequence it is the nation that is presented as the maximal social unit. This is shown graphically in the weather maps visible on the local programs. Usually the full United States is presented with symbols showing the day's national weather pattern. Meanwhile, the world and national news uses world maps as symbols, but treats stories of events in other parts of the world more or less as additions to or spillovers from the national news. Where something is reported from outside the United States, the reporter usually connects the events involved with either American political interests or the vague question of America's standing in the world. Even so, national rather than international news tends to dominate the nationwide broadcasts. Aspects of the local broadcasts foreshadow and provide a bridge to this level of concern, but the nation as such is presented as the maximal structure within which or in relation to which events can become news. At heart, Americans are a little disoriented by world news if it has no American point of reference. This point is evidenced in the fact that the networks feel that maps must be provided to accompany international stories. Otherwise, the American audience has trouble with things like the difference between Hong Kong and Singapore or Zaire and Zambia. Just as American society is almost monolingual from coast to coast, so there is general doubt as to the ontological status of other peoples and countries.

This coast-to-coast image of the United States as the significant context of news events is perhaps the most significant symbolic feature of the TV news. Accident and political history have combined to make this the most complete native model of time and space in popular American culture. People are aware of the four different time zones in the continental United States and the strange infinities of yesterdays and tomorrows that lie beyond the Pacific and Atlantic shores. The frequent visual presentation of the United States map on local weather reports keeps this image in mind. It is even affirmed naturalistically in satellite pictures of the continent.

The time zones have an interesting effect on the six o'clock news because it is known that the end of the all-important "business day" moves from east to west. Events of national importance therefore move from Washington and New York across to Chicago, Denver, and the West Coast. Even the ball scores come in first from the East. The entire daily news cycle completes itself first in the East.

What is more, the names of the four continental time zones reveal some deeper theme in American cosmology. From east to west we have the following zones: Eastern, Central, Mountain, and Pacific. Now here an accident between culture and nature has produced an inconsistency. Logically the sequence "Eastern, Central . . . " should be followed by "Western." But the continent is too wide, so what should be the "Western" time zone has been split into two zones, "Mountain" and "Pacific," named for aspects of nature. Mountains are features of the natural landscape and "Pacific" is the name of an ocean. This terminological adjustment reveals an underlying opposition between East as culture and West as nature. This opposition is further marked by frequent reference to Eastern Standard time as opposed to all others. Standards, of course, are cultural rules which set human life off from nature.

Another accident of geography and history gives added force to this opposition. As I have shown, the events of the daily news cycle move from east to west, just as "civilization" was brought from the East to the frontier back when "the West was won." On the other hand, if we ask, "What moves from west to east in the daily news?" we get a very graphic answer: storm tracks, cloud patterns, and winds. All of these are aspects of nature. And if the news of cultural activities is always one hour behind as we move from east to west across the time zones, then the weather is just the opposite, except that the same regions are about a day behind each other in weather as one moves from west to east. I need not add that a day is a more "natural" category of time than is an hour.

At first it would seem that this contrast between east and west as one between culture and nature is one of mere coincidence. But a second look shows that it is very deeply embedded in American cosmology. Not all cultural developments move from east to west. Some start in California and find their way across to the Atlantic. But what kinds of cultural innovations are these? Surf boards, Hollywood, Disneyland, nude bathing, pet cemeteries, cult groups, dune buggies, hang gliding, and other bizarre or "play" aspects of culture in the sun. Americans expect West Coast culture patterns to have a permanent holiday atmosphere, perhaps even the terror of a holiday gone wrong—but usually something more laughable. Californian culture amounts to cultural plays on nature. It is like the case of the Californian pet cemetery in which the humans want to be buried beside their animals. One bereaved pet owner replied to a news interviewer, "Why should I want to buried with humans?" According to this analysis, the man was absolutely right—why should anyone in California want to be buried with humans? They belong to the fringe of culture.

News, Sports, and Weather

This same opposition between culture and nature is played out another way in the local segments of the six o'clock news. The sequence of news-sports-weather is one which moves from events at the center of culture, that is, problems of government, business, religion, and so on, to athletic competitions, which display cultural plays on natural abilities. Finally the sequence ends with a concern over those aspects of nature that are not subject to human control. Along with this sequence we often get an older man in a business suit reading the news. Next comes a younger man in a sports coat—often of many colors—who reads the sports. Then finally we get the "weather girl" whose garments are often supplied by a local clothing shop. She is there to look pretty and model fashion as well as to give the weather report.

So in going from the news to the weather we repeat what happens in going from east to west. Culture meets nature, with sports as a logical mediator between the two. That women should be identified with the nature side of the news program should not surprise anyone familiar with American culture. We have always named storms after women (though that is now changing), and women have been regarded as unpredictable in their moods just as weather itself is changeable. American males, on the other hand, are assumed, despite all evidence to the contrary, to be rational. They would be more akin to culture and its order than to nature and its disorder.

Just to underline the consistency of these representations, a recent report on CBS radio gave the account of how some animals in a city zoo had been named after the news, sports, and weather. The animals were gnus, and so it was all an elaborate pun. The father gnu was "Sports," the mother gnu was "Weather," and two baby gnus born to them in captivity were named "News" and "Bulletin."

Of course there are exceptions to the patterns identified here. Certain local networks place the weather ahead of the sports. Also, there are more women being employed in the news segment of the programs. This is usually done in a kind of conversational presentation of the news which I think of as the Dick and Jane version. It is conspicuously aimed at balancing women's rights. The variation in the sports-weather sequence is harder to explain. One program director said his station gave the weather first because research had shown that their audience gave weather a higher interest priority than sports. He also claimed that new technology allows us to know the weather more immediately now and that as a result there is no reason to wait until the end of the show for the final predictions from the weather bureau. Thanks to technology, it seems we even know the future sooner!

These rationalistic arguments for the change in sequence do not seem very convincing to me. The rival networks in my area stick to the news-sports-weather sequence and claim that reversals are done merely to be different. My own view is that this variation is not very significant. The native model centers on an opposition between culture and nature. This is represented by the news versus the weather. Sports, in which culture plays with nature, mediates this opposition. It is ritually appropriate to have the sports in the middle, but another reading of the text is to oppose seriousness and play and put the "serious" events of both culture and nature back to back as completing the serious issues of the day and then go on to the play aspects, which happens to resolve the culture/nature opposition. This in a way parallels the work-leisure time dichotomy, which has become more and more prominent as Americans find themselves more consciously using leisure-time activities to identify themselves with various life-style patterns present in the mass culture. Since the work-leisure dichotomy cross-cuts the culture-nature dichotomy the two alternative formats are roughly equivalent.

Conclusion

I have tried to treat TV news formats as expressions of American cosmology. The grounds for doing this are that events must be interpreted before they can be recognized and reported as news. It has been shown that regular features of the TV news format symbolize and affirm some very deeply held beliefs about the larger coordinates of American cultural reality. The news cosmology expressed through the ritual format of news broadcasting appears to me to be one which emphasizes the monocultural, "sea-to-shining-sea" self-image of American society. Other parts of the world are almost off the edge of this nearly flat-earth view. Most curious is the fact that a series of representations concerning news, weather, and sports, as well as male and female, and the time zones, all fit together in such a way as to speak of the conquest of nature by culture and to indicate the public dominance of men over women. The reassuring stability of this framework might be indicated by the fact that we absorb the most unsettling news of the day right along with our heaviest meal.

In the jargon of the American mass media, those who read the news to us are "anchor persons." Although the details require much further study, I hope that my analysis has made it clear why this is an appropriate metaphor. For those who anchor the news broadcasts are in effect anchoring the flow of events to a stable set of understandings about when and where events can take place in the popular American cosmos.

In the fall term of 1977 students in the contemporary American culture course were asked, as their first assignment, to keep a journal noting manifestations of American attitudes about nature, the human body, and science. They were to draw on lecture and reading material, including Miner's famous article "Body Ritual Among the Nacirema," and other selections from Spradley and Rynkiewich (1975) dealing with American values. Gale Thompson's journal included newspaper and magazine advertisements (not reproduced here) and drew on her everyday experiences as an American native. This account points out certain dualistic and contradictory beliefs and attitudes in the contemporary United States. Similar dualistic beliefs show up in other attitudes toward the human body and toward nature.

5 / American Attitudes Toward the Human Body and the Natural World and American Values: You Can't Improve on Nature and Ain't Modern Science Grand?

Gale Thompson

"Back to nature," "the natural look," "good old Mother Nature," and "naturally!" . . . these phrases used by many Americans would seem to indicate a respect for the natural world, but often our behaviors give evidence of the opposite view. Although the natural is "in," it is also *in*decent and *in*adequate. Similarly, our attitudes about our own bodies at first glance seem to be very positive, while later examination shows them otherwise. The human body is a wonder, a miracle; but it too is indecent and inadequate. This may be an invalid judgment when people are studied individually, but, viewed as a whole society, the "Nacirema" appear to hold the dualistic attitudes of respect and disgust in regard to both the human body and the natural world.

The unreserved openness of the natural world sometimes embarrasses us uptight Americans. For instance, consider the mother whose little girl asked about the dogs she saw copulating in the yard. After trying to avoid the question two or three times, the mother finally said, "The brown dog is sick and the black one's pushing it to the hospital." The indecency of the human body and its natural functions is taught early in life, since nudity is taboo in most American homes. Not all cultures have dressing rooms, locks on bathroom doors, or even bathrooms. We hardly ever ask a host, "Where's the toilet?" but rather "the bathroom," "powder room," "john," or even "the little boys' room." We even say instead, "Where can I freshen up?" or " . . . wash my hands?" Public nudity is met with a series of contradictory attitudes in this country. We publish magazines full of male and female nudes, but we put people in jail for "indecent exposure"; we search for a spot to sunbathe on a nude beach, giving ourselves permission for "exposure under sanction," and then fight for our "rights" when someone complains about public nudity. In another contradictory set of behaviors we pass new laws every day to preserve and protect a certain natural area, while at the same time we allow the rape of our land through such activities as strip-mining of coal. Then we kid ourselves into believing that by using more coal we are conserving natural gas. For each value there exists an opposing one and our great democratic system gives us the freedom to choose our own, the right to defend it, and the obligation to allow others to do the same.

Perhaps our self-consciousness about our bodies is a product of the standards we have set. In a world where we find variations from Twiggy to Elizabeth Taylor and from Woody Allen to Orson Welles, we narrow down the acceptable range to an unrealistic ideal—to the extent that only those beautiful bodies in magazines fit the pattern. Then we look at ourselves in a full-length mirror: too fat, too thin, too short, too tall. No wonder we feel inadequate! How can we resolve this conflict between the ideal and the real? Madison Avenue offers us solutions: contour bras, hairpieces for men and women and even "Buns"—the latest in brief underwear for shaping the male derriere. We're amused at natives of other cultures who wear shells in their lips and bones in their noses, while we spend dollars on just the right garments to bring us closer to the desired ideal, and stand for hours in front of a mirror working to achieve that "natural" look.

With a similar approach we look toward nature, seeking just the right spot to locate a new home, school, office building, or shopping center. Then we bring in the bulldozers, chop down the trees, change the contours, erect the building, and afterward call in the landscape architects to recreate the perfect environment. Admittedly, this is being done less and less today, with attempts being made to preserve the natural site, but there are still those who argue that it's cheaper and more expedient to do it the old way. So we're back to improving on nature with our modern science. In fact, we've gone so far as to "plant" artificial evergreens outside of buildings where the site or the management won't support natural growth. Depending on your point of view this is either one step above or below putting plastic plants *inside* the buildings.

Another of our attempts to conform to the ideal standards is the "landscaping" of our bodies, either building them up or trimming them down, or more drastically, resorting to surgery for psychological rather than purely medical reasons. Since the perfect body is not only one of excellent proportion, but one that gives the appearance of eternal youth, cosmetic surgery is becoming more popular, especially for men. To complement our daily rations of vitamins, minerals, and protein, we trim or build our bodies through training and exercise with a variety of techniques ranging from belly dancing to jogging. But if all this seems too tedious there is always the alternative through modern technology of becoming the Bionic Person. Then instead of orange juice, a protein shake, and vitamins for breakfast, we could more appropriately have Tang and artificial eggs and bacon. Does this begin to sound like the familiar old argument of chemical fertilizers vs. the compost heap? Can we improve on nature? Is modern science that great?

In looking further at the way we regard our bodies and the natural world, two other beliefs become apparent. While some of us are diligently caring for our bodies, others of us are nonchalantly ignoring them or perhaps consciously pushing them to the limits. There seems to be a feeling that the human body is the source of infinite energy and that regardless of how we treat it, with lack of sleep, improper nutrition, and little or no exercise, it will keep on functioning to meet all our needs and expectations. There is a parallel to this attitude in the way we treat our natural resources, draining our energy reserves, carelessly using water and land. This behavior seems to have its foundation in the history of America, dating back to the time two hundred years ago when the vastness of the new territory was more than people could comprehend. It's the same way we look at outer space today. In a world this big how could we run out of anything? There's always more "where that came from." Closely related to this is the idea that we and our land are indestructible. No matter how much we smoke or drink or take drugs, somehow we will survive. Our innate belief in the possibility of our own immortality seems to be the basis for this attitude. After all, the scientists can't even agree! Some say that the greenhouse effect will cause the world to overheat; some say the polluted ozone will block the sun's rays and the world will cool off, and some don't believe either will happen . . . so why worry? Some other scientist will discover a way to solve the problem and save us all in the nick of time like our favorite film and television heroes. Perhaps we can thank (or blame) the media for fostering our belief in our indestructibility. If by some chance we do get sick enough to die, those same scientists can just put us on ice and we can wait for thawing until they discover a cure. The cryogenic age, the age of immortality.

Meanwhile we can enjoy living the good life here in America, where we can "get away from it all" and "commune with nature" with all the comforts of home. Or we can bring nature to us, courtesy of hotel chains that provide living plants in every guest room. Here we can change into our bathing suits in the privacy of our own bathroom (decently), and go for a swim. Later we can dress up our inadequate bodies in the latest style and enjoy an evening of whatever hedonistic activities seem most desirable, always remembering that we are fountains of infinite energy and virtually indestructible! At some resort hotels we can swim in clear blue pools of fresh, chemically treated water, while looking out of glass domes at the beach and, beyond it, the ocean. We're not swimming there, of course, because just when we thought it was safe to go in the water, Hollywood showed us the big shark out there, even though it's only a hunk of motorized plastic. We're not ready for the cryovats yet . . . but ain't science wonderful?

The structural similarities of *Star Wars* and *The Wizard of Oz* make them ideal popular cultural products for illustrating Lévi-Strauss's structuralism and Bettelheim's neo-Freudian analysis of folk fairy tales. Here is a detailed treatment that extends structural analysis to the *Star Wars* sequel, *The Empire Strikes Back* (*TESB*). In the next selection, I offer a different kind of interpretation of *TESB*, focusing not on its structural elements, but on the changing currents in American society that made the second film, like its parent, a movie for its time, and a tremendous commercial success.

6 / **Structural and Psychological Analysis of Popular American Fantasy Films**

Conrad Phillip Kottak

One of the main assumptions of anthropologists who study contemporary American culture is that techniques originally developed to analyze cultural features of small-scale societies can also be applied to industrial societies like our own. For example, the structural analysis that Claude Lévi-Strauss (1967) has applied to the myths of Native Americans and other nonindustrial populations can also be used with contemporary forms of narrative expression. People who live in small, unspecialized societies share a common fund of knowledge, acquired through enculturation. A rudimentary division of labor may, for example, assign the role of myth-teller to men, in which case expertise in myth will be shared by many or most men in that society. Or, knowledge of a culture's narratives may be widely shared among most adults.

Americans, too, share knowledge through our common enculturation. We learn national myths in elementary school; every child knows what kind of tree George Washington chopped down, and what he said when asked about his mischievous deed. But our most widely shared experiences with narratives come through the mass media, particularly television. The average American watches several hours each day. (The popularity of the microwave oven may be based as much on its resemblance to a television set as on quick cooking in homes where both spouses work. People can have the illusion that they are cooking in their TV.) Consequently, I have found that one of the most effective ways of teaching contemporary college students how to diagram kinship relationships is to use televised situation comedies to illustrate changes in family patterns. I contrast the nuclear families of the 1950s, as depicted on

"Leave It to Beaver," "Ozzie and Harriet," and "Father Knows Best," with the more complex and blended families shown in more recent programs like "The Brady Bunch," "My Three Sons," "Soap," and "Dallas."

I also gather information on the enculturation that today's college student has experienced by asking a series of questions: How many have never been in a Roman Catholic church in the United States, in a Protestant church, in a Jewish temple? The number of students who respond positively to these questions is always much greater than those who have never eaten at McDonald's, or attended a baseball game, rock concert, or Walt Disney movie. Recently I found that all the students in a class of eighty had seen *The Wizard of Oz* (telecast annually for more than a decade) and all but one had seen *Star Wars*. This provided a basis for their general understanding of the following structural analysis of the two films.

Lévi-Strauss, pursuing the study of cultural universals, has argued that there are basic similarities in the way people think all over the world. There are, he contends, universal structures of the human mind or brain, and these show up in the products (including creative expression) of all known cultures. He asserts that people everywhere have a need to classify, to categorize their own experiences and the world around them, and that one of the most common ways of doing this is by binary opposition. Thus, things and qualities that in nature are continuous, that really differ in degree rather than in kind, are treated as discrete and opposite. For example, the continuum that runs from tall to short is dichotomized simply into "tall" and "short." Good and evil are similar polarizations of contrasts that are not actually so clear-cut.

Lévi-Strauss has applied his assumptions about classification and binary opposition to myths and folk tales, showing that these narratives are often made up of simple building blocks, elementary structures, or "mythemes." Examining the myths of different cultures, Lévi-Strauss shows that one tale can be converted into another through a series of simple operations; for example, converting the positive element of one myth into its negative, reversing the order of the elements, replacing a male hero with a female, and preserving or repeating certain key elements. As a result of such operations, two apparently dissimilar myths can be shown to be variations on a common structure, i.e., to be transformations of each other. One example is Lévi-Strauss's (1967) analysis of the "Cinderella" story, a widespread tale whose essential elements vary from one culture to its neighbor. (To understand that fairy tales have a series of different versions, think of the several endings of "The

Three Little Pigs" and of "Little Red Riding Hood.") Eventually, after a sufficient number of reversals, oppositions, and negations as the tale is told, retold, and incorporated into the traditions of successive societies, "Cinderella" becomes "Ash Boy," with a series of simple contrasts related to the change in hero's gender. (For other examples of this, see Lévi-Strauss 1969; but see also Marvin Harris's [1979] important critique of the limitations of structural analysis.)

I will argue in this essay that *Star Wars* is a systematic structural transformation of *The Wizard of Oz*, even more obviously (to American natives) than "Ash Boy" is a transformation of "Cinderella." To round out the following analysis and extend it a bit farther, to *Star Wars*'s sequel, *The Empire Strikes Back* (*TESB*), I will also use techniques developed by neo-Freudian psychoanalyst Bruno Bettelheim.

In his book *The Uses of Enchantment: The Meaning and Importance of Fairy Tales*, Bettelheim (1975) urges parents to read or tell folk fairy tales to their children. He chides American parents and librarians for pushing children to read "realistic" stories, which often are dull, too complex, and psychologically empty for children. Folk fairy tales, in contrast, permit children to identify with heroes who win out in the end, offering confidence that no matter how bad things seem now, they will eventually improve. They offer reassurance that, though small and insignificant now, the child will eventually grow up and achieve independence from parents and siblings.

Related to Lévi-Strauss's focus on binary oppositions is Bettelheim's analysis of how fairy tales permit children to deal with their ambivalent feelings about their parents and siblings. Children both love and hate their parents. Bettelheim tells of one girl who, when scolded and yelled at by her mother, developed the fantasy that a Martian was temporarily inhabiting the mother's body, as an explanation for her change in mood. Fairy tales often split the good and bad aspects of the parent into separate figures of good or evil. Thus in "Cinderella," the mother is split in two, an evil stepmother and a fairy go(o)dmother. And Cinderella's two evil stepsisters disguise the child's hostile and rivalrous feelings toward his or her real siblings. A tale like "Cinderella" permits the child to deal, guiltlessly, with hostile feelings toward parents and siblings, since positive feelings are preserved in the idealized good figure.

According to Bettelheim, it doesn't matter much whether the hero is male or female, since children of both sexes can usually find psychological satisfaction of some sort from a fairy tale. However, male heroes typically slay dragons, giants, or monsters (representing the father) and free princesses from captivity, whereas female characters accomplish something and establish a home of their own.

The contributions of Lévi-Strauss and Bettelheim permit the following analysis of the visual fairy tales that contemporary Americans know best. I cannot say how many of these resemblances were part of a conscious design by *Star Wars*'s writer and director George Lucas and how many were manifestations of a collective unconscious that Lucas shares with us through common enculturation.

The Wizard of Oz and *Star Wars* both begin in arid country, the first in Kansas, the second on the desert planet Tatooine (see table 1). *Star Wars* changes *The Wizard*'s female hero into a boy, Luke Skywalker. As in fairy tales, both heroes have short, common first names, and second names that describe their ambience and activity. Thus Luke, who travels aboard spaceships, is a Skywalker, while Dorothy Gale is swept off to Oz by a cyclone (a gale of wind). Dorothy leaves home with her dog Toto, who is being pursued by, and has managed to escape from, a woman who in Oz becomes the Wicked Witch of the West. Luke follows his own "Two-Two" (R2D2), who is fleeing Darth Vader, the witch's structural equivalent.

Dorothy and Luke both live with an uncle and aunt, but because of the gender change of the hero, the primary relationship is reversed and inverted. Thus Dorothy's relationship with her aunt is primary, warm, and loving, whereas Luke's relationship with his uncle, though primary, is strained and distant. Aunt and uncle are in the tales for the same reason. They represent home (the nuclear family of orientation), which children must eventually leave to make it on their own. Yet, as Bettelheim points out, disguising parents as uncle and aunt establishes social distance; the child can deal with the hero's separation (in *The Wizard of Oz*) or the aunt's and uncle's death (in *Star Wars*) more easily than with the death or separation of the real parents. Furthermore, this permits the child's strong feelings toward his or her real parents to be represented in different, more central characters.

Both films focus on the child's relationship with the parent of the same sex, dividing that parent into three parts. In *The Wizard*, the mother is split into two parts bad and one good: the Wicked Witch of the East, dead at the beginning of the movie; the Wicked Witch of the West, dead at the end; and Glinda, the goodmother, who survives. *Star Wars* reverses the proportion of good and bad, giving Luke a good father (his own), the Jedi knight who is dead at the film's beginning; another good father, Ben Kenobi, who is ambiguously dead when the

TABLE 1 *Star Wars* as a Structural Transformation of *The Wizard of Oz*

Star Wars	*The Wizard of Oz*
Male hero (Luke Skywalker)	Female hero (Dorothy Gale)
Arid Tatooine	Arid Kansas
Luke follows R2D2 R2D2 flees Vader	Dorothy follows Toto Toto flees witch
Luke lives with uncle and aunt Primary relationship with uncle (same sex as hero) Strained, distant relationship with uncle	Dorothy lives with uncle and aunt Primary relationship with aunt (same sex as hero) Warm, close relationship with aunt
Tripartite division of same-sex parent 2 parts good, 1 part bad father Good father dead at beginning Good father dead (?) at end Bad father survives	Tripartite division of same-sex parent 2 parts bad, 1 part good mother Bad mother dead at beginning Bad mother dead at end Good mother survives
Relationship with parent of opposite sex (Princess Leia Organa): Princess is unwilling captive Needle Princess is freed	Relationship with parent of opposite sex (The Wizard of Oz): Wizard makes impossible demands Broomstick Wizard turns out to be sham
Trio of companions: Han Solo, C3PO, Chewbacca	Trio of companions: Scarecrow, Tin Woodman, Cowardly Lion
Minor characters: Jawas Sand People Stormtroopers	Minor characters: Munchkins Apple Trees Flying Monkeys
Settings: Death Star Verdant Tikal (rebel base)	Settings: Witch's castle Emerald City
Conclusion: Luke uses magic to accomplish goal (destroy Death Star)	Conclusion: Dorothy uses magic to accomplish goal (return to Kansas)

movie ends; and a father figure of total evil. It is easy to note the phonetic resemblance of Darth Vader to "dark father." In a *New York Times* interview (May 18, 1980), just before the opening of *The Empire Strikes Back*, Lucas claimed that he chose "Darth Vader" because it sounded like both "dark father" and "deathwater." As the goodmother third survives *The Wizard of Oz*, the badfather third lives on after *Star Wars*, to strike back in the sequel.

The child's relationship with the parent of opposite sex is also represented in the two films. Dorothy's father figure is the Wizard of Oz, initially a terrifying figure, later proved to be a fake. Bettelheim notes that the typical fairy tale father is either disguised as a monster or giant, or else (when preserved as a human) is weak, distant, or ineffective. Children wonder why Cinderella's father permits her to be treated badly by her stepmother and siblings, why the father of Hansel and Gretel doesn't throw out his new wife instead of his children, and why Mr. White, Snow White's father, doesn't tell the queen she's too narcissistic. Dorothy counts on the wizard to save her, finds that he is posing seemingly impossible demands, achieves significantly on her own, and no longer relies on a father who offers no more than she herself possesses.

Luke's mother figure is Princess Leia Organa. As Bettelheim notes, early-Oedipal boys commonly fantasize their mothers to be unwilling captives of their fathers, and fairy tales frequently disguise mothers as princesses, whose freedom the boy-hero must obtain. In graphic Freudian imagery, Darth Vader threatens Princess Leia with a needle the size of the witch's broomstick. By the end of the film, Luke has freed Leia, vanquished Vader, and the princess seems destined to become Ms. Organa-Skywalker.

There are other striking parallels in the structures of the two films. Fairy-tale heroes are often accompanied on their adventures by secondary characters who personify virtues needed in a successful quest. Dorothy takes along wisdom (the Scarecrow), love (the Tin Woodman), and courage (the Cowardly Lion). *Star Wars* includes a structurally equivalent trio—Han Solo, C3PO, and Chewbacca—but their association with particular qualities is not as precise. The minor characters are also structurally parallel—Munchkins and Jawas, Apple Trees and Sand People, Flying Monkeys and Stormtroopers. And compare settings—the witch's castle and the Death Star, the Emerald City and the rebel base (filmed at Tikal, in verdant Guatemala). The endings are also parallel. Luke accomplishes his objective on his own, using the Force (Oceanian *mana*, magical power). Dorothy's aim is to return to Kansas; that she does, tapping her shoes together, drawing on the Force in her ruby slippers.

I have previously argued (Kottak 1978*b*) that these resemblances help explain *Star Wars*'s huge success. It is likely that all successful cultural products blend old and new, draw on familiar themes, rearrange them in novel ways, and thus win a lasting place in the imaginations of whatever culture creates or accepts them. *Star Wars* successfully used old cultural themes in novel ways, and it came along at an optimistic time, a time for heroes, in American culture. It drew on *the* American fairy tale, one that had been available in book form since the turn of the century. The movie version of *The Wizard of Oz* was not immediately successful when it was released in 1939, the same year as *Gone with the Wind*, which found a much larger immediate audience. *The Wizard*'s popularity had to await happier years, and annual telecasting that brings it into every home. Our familiarity with this narrative therefore comes more from television than from movies.

In 1980 *Star Wars*'s sequel, *TESB*, rivaled the success of the 1977 film. Was *TESB* a structural transformation of any previous work of American culture? I considered *Gone with the Wind* and *Casablanca*, two popular old films with stereotyped and memorable characters, but I found only minor parallels. I soon did discover the source of *TESB*'s structure in a previous film—*Star Wars* itself. In the case of *Star Wars*'s transformation of *The Wizard*, most of the structural contrasts were simply those that logically followed from the change from female to male hero. There were only a few structural inversions (two parts bad to two parts good, and goodmother versus badfather lives on). *TESB* transforms *Star Wars* differently, but also through a series of simple operations. Rather than the gender change, there is a partial shift in perspective from young hero to old villain, and a series of elements are converted into their opposites. (Note that being opposite or inverted is not the same as a gender change, since a male is not really the opposite of a female, though our culture sometimes considers them as such.) *TESB* is a negation, accomplished through a fairly consistent series of simple structural inversions of elements of the original. The trilogy awaits conclusion in the next film, to be, in dialectical terms, the negation of the negation, the synthesis of the thesis (*Star Wars*) and the antithesis (*TESB*).

But let us look at *TESB*. There are some important general oppositions. First, in *Star Wars* Luke was preoccupied with freeing the mother figure; here he is absorbed in relationships with father figures. Second, Darth Vader dominates the second film much as Luke dominated the first. Third, Vader becomes more human in *TESB* (for example, through a quick shot of his pink head as his helmet is lowered, and through his emotional invitation to his son to join him in ruling the galaxy). Luke simultaneously becomes less clearly good as he flirts with the dark side of the Force (killing Vader's image which turns out to have his face, and becoming partially bionic like his father). Note that Luke is shown upside down several times in *TESB*, symbolically suggesting the overall turnaround. That inversion (table 2) is marked from the very beginning, the opening shot of a tall, spindly, dark, imperial robot landing on a cold, wet planet—the opposite of the *Star Wars* opening, in which a short, squat, light, rebel robot (R2D2) landed on a hot, dry planet. Luke is almost immediately hung up by an ice creature, as if to say at once that everything here is upside down. In *Star Wars* Luke saw an image of the mother figure that propelled him on his adventures. Here an image of Ben (an aspect of the father) serves the same function. In the first film Luke was gradually joined by a party of companions. Here they split up.

Yoda's inclusion offers structural balance to several elements of the *Star Wars* plot. As a dramatic presence he replaces Ben Kenobi. As an intelligent, alien, cartoonlike nonhuman, he substitutes for the aliens in the celebrated cantina scene of the first film. Most important, he fills the gap in the tripartite division of the father into two parts good and one bad when Darth Vader reveals himself to be Luke's real father, previously thought dead. In addition to his role as triple structural equivalent, Yoda is, in Lévi-Strauss's terms, a mediating figure par excellance, since he links Luke, Ben, and Darth Vader, having taught all three. He also links the positive and negative sides of the Force, which his students have used for good and for evil.

The two films proceed through a similar order of events, which are often simple inversions of the original structure (e.g., good father becomes bad father, dead father becomes live father). Occasionally the parallel parts of the plots are similar in content. Thus, in *Star Wars* the Millenium Falcon emerged from hyperspace amid the rock fragments of Alderan's destruction, whereas in *TESB*, unable to enter hyperspace (an inversion), the Falcon must dodge other rocks—asteroids. And there are some structural equivalents. The giant worm that swallows the Falcon in *TESB* replaces the garbage snake-monster of *Star Wars*. The heroes eventually converge on the City in the Clouds, which takes the place of

TABLE 2 *The Empire Strikes Back (TESB)* as a Structural Negation of *Star Wars*

The Empire Strikes Back	*Star Wars*
Tall, spindly, dark imperial robot lands on cold, wet planet	Short, squat, light rebel robot lands on hot, dry planet
Image of Ben starts Luke on adventures	Image of Leia starts Luke on adventures
Luke separates from companions	Luke is joined by companions
Vader gradually disposes of assistants	Luke gradually acquires assistants
Yoda replaces . . .	Ben, as dramatic presence
Yoda replaces . . .	Cantina scene, as nonhuman
Yoda replaces . . . (real father alive and evil)	Luke's dead good father (real father dead and good)
Luke experiences dark side of The Force	Luke experiments with good side of The Force
No hyperspace drive—rocks	Out of hyperspace—rocks
Giant worm	Garbage snake
City in clouds	Death Star
Han departs prematurely; doesn't return	Han departs prematurely; returns
Lando Calrissian replaces . . .	Han Solo
Luke's real father alive and evil (told late in film)	Luke's real father dead and good (told early in film)
Young, good figure battles evil, gives up; undergoes (emotional) transformation	Old, good figure battles evil, gives up; undergoes (spiritual) transformation
Luke rejects bad father's life	Luke rejects good father's death
Bad father deprives Luke of light saber	Good father gives Luke light saber
Leia saves Luke	Luke saves Leia
Luke identifies with bad father (telepathically and bionically)	Luke identifies with good father (telepathically and spiritually)
Vader wins	Luke wins
Luke (temporarily defeated) leaves in ship	Vader (temporarily defeated) leaves in ship
Companions split	Companions assemble
Vader isolated	Luke social

the Death Star. They encounter Lando Calrissian, who eventually substitutes for Han Solo.

Many of the most striking oppositions come at the end. For example, in *Star Wars* Han's premature departure was resolved; he returned to help Luke in the final battle. But here his departure is unresolved; he provides no support for Luke's climactic battle. In *Star Wars*, the old good figure (Ben) battled evil, gave up, but underwent a transformation, to become spirit; here the young good figure (Luke) battles evil, gives up, and undergoes an emotional transformation by confronting his parentage. After *Star Wars*'s duel, Luke yelled "No!" rejecting the good father's death. Here he yells "No!" rejecting the bad father's life. In *Star Wars* Ben presented Luke with a light saber (symbolic identification with the father), but here Vader slashes off Luke's saber, along with his hand. In *Star Wars* Luke used the Force to destroy evil; here he uses the Force to call Leia. He saved her in *Star Wars*; now she saves him.

In *Star Wars* Luke partially mastered the good side of the Force and became spiritually like the goodfather. In *TESB* Luke flirts with the dark side of the Force and by the film's end has become mechanically like his bionic badfather. There he drew on Ben's spiritual presence and telepathic contact to destroy the Death Star; here he is telepathically touched by the darkfather. At the end of *Star Wars* Vader left in a ship; here Luke does. There the companions came together; here they split up.

Star Wars concluded with Luke's triumph against seemingly impossible odds. Here he doesn't win; he simply escapes, like Vader in the original, to strike back in the next episode. At the end of *Star Wars* Luke Skywalker and the loyal companions he had gradually assembled during the film celebrated together. But by the end of *TESB*, Darth Vader has systematically disposed of his three assistants, admirals who have failed him in reaching his objectives. Vader's isolation and alienation from other people stands in sharp contrast to Luke's identity as a social person. Thus, at the end of *TESB* Vader stands alone, without a son, as Luke was without a father at the beginning of *Star Wars*. My prediction for the trilogy's conclusion, for the negation of the negation: both Luke and Vader will be social and together, reunited as reformed Jedi and equal son.

Many other elements could be mentioned to demonstrate that *Star Wars* and *TESB* are structural transformations of one another, but I leave that to other analysts. My intent here has been to use three extremely popular products of the electronic mass media to show that techniques developed to analyze the expressive forms of simpler societies can also be applied to our own. Use of structural analysis and of Bettelheim's research on the nature of fairy tales and their psychological significance helps uncover the enduring cultural meaning of these films. It also confirms Lévi-Strauss's finding, from the myths of non-Western societies, that narratives that on the surface seem very different may share a deeper structure that is not very different at all. *Star Wars* used a gender contrast, structural equivalents, and a few inversions to transform *The Wizard of Oz. TESB* uses inversion, structural equivalents, and reversal of outcome to transform *Star Wars*.

Consider this final structural twist, a last major similarity between the *Star Wars* films and *Oz*. The literary sequel to *The Wizard of Oz, The Land of Oz,* also patterned itself on the structure of the original, using such simple operations as a gender contrast and substitution of structural equivalents to accomplish the transformation. Instead of Dorothy, *The Land of Oz* has a male hero, Tip. A sawhorse replaces Toto, and a trio composed of Pumpkinhead, Wogglebug, and sofa-creature stand in for Dorothy's three companions. However, in one of the most psychologically curious outcomes in children's literature, the hero doesn't save the princess, but turns into a princess himself (into Ozma, the "girlish" ruler of Oz). That transformation probably explains why the *Oz* books never underwent the kind of movie serialization that the *Star Wars* story is undergoing. It shows as well that even similar structures can't produce success when, in content, fundamental cultural assumptions are violated. Boys, after all, should turn not into girls, but into men.

In the previous selection, I offered a detailed structural analysis of *The Empire Strikes Back* (*TESB*), showing its relationship to *Star Wars*, itself a structural transformation of *The Wizard of Oz*. Here I try to place *TESB* in its social context, and I stress the film's content rather than its structure. I point out that the film's social psychological significance draws on enculturative experiences shared by millions of Americans; I suggest that successful films are the ones to which we can most easily and imaginatively free-associate. Here I relate a hugely successful cultural product to many other aspects of American culture: sex-role stereotyping, changing family patterns, values, and even baseball.

7 / **The Father Strikes Back**

Conrad Phillip Kottak

May 21, 1980, opening day for *The Empire Strikes Back*, a national cultural event. At the Americana theater in suburban Detroit people began lining up at 9 A.M. for the first showing (12:25). The crowd for the 2:55 show, which I attended, was near the theater's capacity of seventeen hundred. The line reminded me of queuing behavior at Walt Disney World, the waits for the Haunted Mansion and Space Mountain. Here, however, the carnival ride in American culture required no trip to Florida or California; one could find fantasy-escape at a local movie house. Having read favorable reviews, I suspected that people would see *The Empire Strikes Back* (*TESB*) over and over again, as they had seen *Star Wars*, particularly since they had been preprogrammed to like the sequel by the original's success. American natives would return to this film for some of the same reasons they revisit Disney's Magic Kingdom: the simultaneous presence of past, present, future, fantasy, and the familiar in a cultural product that allows vicarious escape.

I was right. The newer film rivaled the success of the original. A significant media event such as an extremely popular film is not a thing in itself. Critics who look only at the product miss what is most significant to the anthropologist, the interaction between the cultural product and the society that receives it (and whether it is received at all). Key beliefs in American culture have been in the propriety of realism over fantasy and in our technology's eventual ability to solve any problem. These beliefs, which cannot be scientifically demonstrated to be either true or false, show up clearly in commentary on different genres of film. Thus, reviewers typically praise special effects in space and fantasy films, as if (after *The Black Hole* and *Battlestar Galactica*) this could explain a film's success. And they see "realistic" films like *Kramer versus Kramer*, *Breaking Away*, and *Ordinary People* as addressing social changes in our society, while totally missing manifestations of exactly the same themes in films like *TESB*. The very simplicity of an effective visual fairy tale, one of the main reasons for its generalized and enduring cultural significance, is sometimes labeled a defect. Although fantasy-escape may be okay for children, it is not easy to encourage in a society that measures the worth of men, women, teenagers, and the elderly by the amount they earn on the extradomestic work force.

Yet millions of people did appreciate *TESB*. The film has a faster pace and greater complexity than *Star Wars*. To comprehend it, you must see it more than once. And, with successive viewings individuals bring their own associations to and into it, as, according to Bruno Bettelheim (1975), children do with folk fairy tales. Anthropologists focus not on the special meaning that a cultural phenomenon has to a particular person, but on *common* reactions and perceptions, conscious and unconscious, that people share through enculturation, having been raised in the same culture. Like children listening to fairy tales, Americans work through unconscious material when we participate in such an event, but much of our free association works on material that we share because of our common cultural heritage. The more there is to associate to, the more successful the film will be, as *Star Wars*, with its references to numerous movies, novels, television programs, comic books, and other previous statements of American culture, demonstrated. *TESB* evoked a similar range of associations. For example, every native I queried about *TESB*'s advertising poster that showed Han Solo and Princess Leia against a bright orange background immediately associated it with the famous poster for *Gone with the Wind*.

As an anthropologist I often teach about unconscious and undetected themes and associations in many cultures, including our own. I am also an incorrigible punster. Looking at the advertising for *TESB*, in which Vader's face mask was prominent, and considering the title, I free-associated *TESB* to the *umpire strike* of the summer of 1979. Darth Vader, the man behind the plate, looked like a baseball umpire. Could baseball and the *Star Wars* films be associated metaphors for themes in American culture? Football has been seen as metaphorizing competition, aggression, specialization, and group coordination in American life. Baseball focuses more on individuals, and fans identify with the flaws, heroics, and statistics of particular players. I could not imagine *Star*

Wars characters playing professional football, but it was easy to see them on the baseball diamond. I even composed a team using the eight main *Star Wars* characters. Luke was pitcher, Leia catcher, Han Solo played first base, C3PO second, R2D2 shortstop, Chewbacca third, Darth Vader left ("sinister") field, and Ben Kenobi right (as in justice). Center field was vacant.

Was my association between the *Star Wars* films and baseball idiosyncratic or cultural? Would other natives assign the characters similarly? I distributed questionnaires to seventy-six students (who had seen *Star Wars*) in two undergraduate anthropology courses. They were asked to assign the nine positions to the eight main *Star Wars* characters. *Umpire* was not listed, and most of the students had not yet seen *TESB*, which came to Ann Arbor a month later, but a few spontaneously connected Darth Vader with that position of authority, which often receives fans' hostility.

Since there were only eight characters to be assigned to nine positions, the responses suggested which positions Americans consider most critical. Most frequently assigned were pitcher (seventy-six), catcher (seventy-four), and first base (seventy-three). Least frequently filled were right field (forty-nine), center (fifty-six), and second base (sixty-one). Although the students' answers didn't always match my own, they were not assigning positions at random. Some collective associations *were* being revealed. Darth Vader (twenty) and Luke (eighteen) were virtually tied for pitcher, confirming their roles as the main characters. R2D2 won two positions: shortstop (fourteen), and catcher (thirty) overwhelmingly. Princess Leia won right field (eighteen) and was runner-up at second base (with twelve votes to C3PO's fifteen). Americans' prejudice against females' playing baseball showed up here, as these were considered two of the least important positions. Chewbacca and Ben Kenobi received ten votes each in left field, and Han barely won center with twelve votes, versus nine each for R2D2 and C3PO.

The most obvious conclusion from the questionnaire was that students were sufficiently familiar with both baseball and *Star Wars* to feel comfortable with the questionnaire; no one had to ask me for clarification. These were young respondents (average age about twenty) whose perceptions of baseball positions may reflect more their own recent playing experience than seeing telecasts of the game. Professional right fielders and second basemen must be good players, since professional baseball employs talented left-handed hitters as well as the right-handed batters who predominate in school play and make left field the critical outfield position.

The students disagreed with me most about three assignments. Only three made Leia catcher, a position I thought she deserved as part of a pair with Luke (as pitcher). Only eight put Darth Vader in left, and seven put him in right; one told me that she associated Vader with "right-wing." I had placed Ben in right field, but only three students did, perhaps because of the previous association, but most probably because he was too significant a character for the least-favored position.

However, the students did agree that Luke and Vader were the main characters, and this is even truer in *TESB* than in *Star Wars*. My conversations with children who have seen *TESB* convince me that they still identify with Luke, as they must with a fairy tale hero. Yet the characters are more complex in *TESB*, and the film is Vader's as much as Luke's. *Star Wars* symbolized the early-Oedipal boy's wish to possess the mother, seen as the unwilling captive of the father, whom the boy wishes to remove. As in numerous fairy tales, the mother is disguised as a princess; by the end of *Star Wars* Luke had vanquished the dark father, and the princess seemed his eventual reward. *TESB*, however, focuses on the son-father relationship, and Luke doesn't seem to mind that the princess is favoring the older-brother figure (Han Solo). Luke has his masculine identification to work out; the struggle with the father is the critical issue.

According to Bettelheim, such obvious centrality of the father-son relationship is rare in fairy tales. But *TESB*'s preoccupation with the father also reflects larger forces in American culture, particularly changes in parental roles related to rising female extradomestic employment and greater male participation in child rearing. Bettelheim found that fairy-tale fathers were almost always disguised as giants, dragons, and monsters, whereas mothers were preserved as human and kept around the home, where they were in real life. The good and bad qualities of the mother (but rarely of the father) were often split into separate figures of good or evil, like Cinderella's stepmother and fairy godmother. Both *Star Wars* films do split the father image in exactly this way, and this important contrast with older tales is symbolic of our changing family relationships. Issues of father-son attachment and separation have also been central to many recent films—for example, *Kramer versus Kramer* and *Ordinary People*.

In *Star Wars* Luke is mesmerized by Leia's (the mother figure) image, which starts him on his adventures, whereas in *TESB*, an image of Ben Kenobi (representing the good father) directs Luke to pursue his study of the Force. On the marsh planet Dagobah, Luke meets Yoda, another aspect of the father. The inclusion of Yoda preserves the original film's tripartite division of the father into two parts good and one part bad, since Darth Vader turns out to be Luke's actual father. Though the good father we thought dead is alive and evil, Yoda is an adequate substitute.

The main appeal of *TESB* is to the child leaving the Oedipal period for latency and the residue of those feelings that remain in teenagers and adults. The resolution of the Oedipal period is represented in Luke's climactic battle with Vader, in which Luke is symbolically castrated, losing his light saber and his hand. Thereafter there is partial identification with the aggressor as Luke gives up, plunges into an abyss, and later acknowledges his paternity as Vader probes him telepathically. By the end of the movie Luke has become more like his father. He is now part mechanical, the recipient of a bionic hand.

Greater maturity and achievement permit Luke to accept an evil father as his own. He can dispense with the fiction that his father is an idealized dead knight. *Star Wars* confronted the child's separation anxiety, wrenching Luke from home (disguising parents as uncle and aunt). Luke's problem in *TESB* is just the opposite: he must fend off an all-powerful father who has returned to deny his independence, a bionic man of evil who can penetrate to the child's very unconscious with his force.

Latency is a period of psychological development during which aggressive and sexual impulses are (relatively) subdued. Luke struggles, confronting the dubious aspects of self and father. Battered, bruised, he has grown and learned by the end of the film, but Luke remains incomplete. His full personality integration awaits his third film; here he (temporarily) gives in to the renewal of parental authority.

Although the critical developmental periods in *TESB* may be late-Oedipal and latency, there is something for younger children, too. Early-Oedipal period boys, for example, resent the romance between Han Solo, a clear adolescent, and Leia, who should be appreciating Luke's growth. Like older brothers in fairy tales, Han's fate is an unhappy one. Darth Vader, though preoccupied with Luke, pauses long enough to turn Han into something that resembles stone, which children may interpret as punishment for inappropriate flirtation with the mother figure. Why else would little boys I have queried say that they like Boba Fett, the bounty hunter who takes Han away?

Another message: a boy can still rely on the mother during his struggle with the father. Leia saves Luke here, just the reverse of *Star Wars*.

If Luke must temporarily submit, Darth Vader's fate is far unhappier, the opposite of Luke's achievement in *Star Wars*. By the end of *TESB*, Vader is alone. He has nothing but the Force, which he can use only destructively—for example, to kill three generals. How different from Luke, who can rely on loyal companions.

Let me return now to the metaphor of baseball, which has also been associated with the father-child relationship in traditional American culture. A new team is necessary for *TESB*. Luke remains pitcher, with Leia moving to short. I defer to the students and let R2D2 catch, especially since he is paired with Luke throughout most of the movie. Lando Calrissian replaces Han at first; Han has been demoted to left field because of his flirtation with Leia. C3PO remains at second and Chewbacca at third. Ben stays in right, and Yoda moves into center, because, as one informant put it, he teaches the Force, which people can use for good or evil. But where is Darth Vader? Off the team altogether, behind the plate, as enduring authority, umpire.

A thousand films are made in a decade, but few become culturally significant and therefore worthy of anthropological attention. The ones that have special meaning to millions of people are well made, and they appeal to our need to think, to imagine, and to deal with generalized problems of human existence. They succeed best when their particular messages are also appropriate for specific social circumstances. Both *Star Wars* and *TESB* were movies for their times, but the times were different. *Star Wars* gave us a hero, undiluted hope, fun, fantasy, and the future. That was in 1977, before the dissipation of the "bicentennial spirit," before the gasoline shortages, the energy crisis, inflation, Iran, Afghanistan, and recession. By the summer of 1980 we had moved from the short-lived period of optimism that followed the Vietnam and Watergate years to an age of uncertainty.

A similar uncertainty pervades *TESB*. The dilemmas are not resolved. We must await the third film to find out if Han is freed, if Luke masters the Force and surpasses Vader, if Luke and Leia are united. *TESB*'s ending is not really happy, not really sad. This time ordinary people do not triumph, nor do they lose. Things are suspended, but there is a ray of hope. After all, Luke does escape against seemingly impossible odds.

During the summer of 1980 news was also uncertain. Who would be elected president? Did it matter? Would the recession get worse? In contrast to the Johnson-Nixon years, public figures were not seen as good or evil incarnate, but as ambiguous and ordinary (as they really are). Uncertainty and indecisiveness were familiar themes in American culture in 1980. Americans could not look forward to their compatriots attending and triumphing in the summer Olympics. No heroes were in sight. And, if the country had become reconciled to three hundred days of waiting for fifty real hostages to be freed from Iran, it could surely tolerate three more years of captivity for Han Solo.

In this short essay (an early class assignment), Thompson illustrates different approaches to the analysis of myth using fairy tales, myths, or stories popular in contemporary America. She applies Lévi-Strauss's structuralism, Victor Turner's symbolism, and Bettelheim's neo-Freudianism to an award-winning film and a traditional fairy tale. Her conclusion sees both tales in the context of their culture of origin, feudal Europe and ethnic America.

8 / Approaches to the Analysis of Myth, Illustrated by *West Side Story* and "Snow White"

Gale Thompson

> Myth: *mith (Gk mythos) a usually traditional story of ostensibly historical events that serves to unfold a part of the world view of a people or explain a practice, belief, or natural phenomenon.*

Perhaps the secret of the universal appeal of so many myths lies in the origin of the word itself. It derives from the Greek *mythos*, whose root sound is *mu*, which means to mumble, to make sound with the mouth. Mythology then becomes a study of a simple, basic, even primal form of expression. Myths proclaim our common identity as human beings. Their ability to transcend language, culture, history, and religion makes the myths of any country easily accessible to peoples of many countries. A major theme in many myths is the search for the ultimate hero who would, of course, be immortal. The links between myth and religion are apparent when we consider the common meanings in the Chinese *Tao*, the Hindu *Brahma*, and the Judeo-Christian *Word*. "In the beginning was the Word . . . and the Word was made flesh and dwelt among us." The great heroic myth can be divided into eight phases: miraculous birth, initiation, withdrawal, quest or trial, death, descent into the underworld, rebirth, and elevation to divine status. Buddha, Christ, Mohammed—the list is a long one when we consider those who have made this hero's journey, and their stories convey the message that this could be our journey as well. Small wonder that myths have endured for centuries.

While myths can be appreciated and enjoyed simply as stories, they hold far more than just entertainment value. We can peel away the layers of a myth just like those of an onion and find a variety of meanings and messages. Each new interpretation can make the myth more valid and more valuable. The most logical approach is to look at the moral, or face value, of the story itself. *West Side Story*, a tale of mid-twentieth-century modern star-crossed lovers, shows a belief that love conquers hate as animosity between two warring factions subsides, but not until one of the lovers pays the price with his life. Depending on how one views Tony's personality this could be a confirmation that "evil is always punished" or that "nice guys finish last."

Using Lévi-Strauss's approach of structural analysis, we can examine binary oppositions within the same tale to discover some of the cultural themes operating in America during the 1950s. Some examples are easily seen:

Maria/Tony	nature/culture	female/male
female/male	Sharks/Jets	gentle/brusque
life/death		passive/active
		sensitive/tough

The roles of Tony and Maria point out the traditional view of male/female characteristics, since they fit the stereotypes so familiar in that era. The woman belongs in the background, especially in times of danger. Man is the hunter, the warrior, the protector. It would be hard to imagine Maria setting out to avenge the death of her brother. And yet in other cultures or in other times that might be expected.

Looking further it is possible to discover the usual mediating figures in this contemporary Romeo/Juliet tale:

	Mediating Figures	
Italians	Tony/Maria	Puerto Ricans
Jets/Sharks	Tony/Maria	The Establishment
Jets/Sharks	Officer Krupke	The Establishment
teens	Drug Store Man/ seamstress	adults

There is evidence here of the concepts of bilateral kinship, fictive kinship, and the "generation gap" idea so prevalent at the time *West Side Story* became an award-winning film. Obviously, it had a lot to relate to us about our values.

With our current efforts toward changing the role of women in our society and the new emphasis on being able to communicate with others, except for its nostalgia, we may not find *West Side Story* to have such great appeal in America today, if we consider it only in terms of this interpretation.

Another method; another myth. Using Victor Turner's (1974) concepts of ritual and liminality, the tale of Snow White becomes more than just a story about a pretty little girl who lives happily ever after with the local handsome prince. As for rituals, like many fairy tales this one has a "chant" that is repeated seven times in the three versions selected at random from the children's department in the Ann Arbor public library:

> "Mirror, mirror on the wall, Who is the fairest one of all?"
> And each day the mirror replied, "Queen, you are fairest of them all."

Of course, the vain queen got this answer only one time. As Snow White grew up the answer changed to:

> "Queen, though you are fair, 'tis true, Snow White is fairer far than you."

Then another series of rituals begins as the queen tries three times to do away with her competition. One earlier attempt had succeeded, she thought, as the huntsman had brought back the heart of a boar to "prove" that he had obeyed the queen's command.

The dwarfs also have a rituallike chanting episode when they discover that someone has invaded their home and they ask, "Who's been sitting in my little chair? . . . using my little spoon?" . . . etc. Disney's film expands their roles to include chants about whistling while you work, scrubbing clean, and "Heigh-Ho"-ing off to work each day. Our Protestant work ethic comes through loud and clear in his version, as does the Puritan "cleanliness is next to godliness."

Beyond the surface it is interesting to note the same color symbolism that Turner illustrated in recounting Ndembu rituals, only this time the colors are reversed. Consider the following:

> It was the middle of winter, and flakes of snow were falling about the royal castle of a far-off land. By her window, the queen of the country sat sewing at her embroidery frame of fine black ebony. As she looked out upon the snow, she pricked her finger and three drops of blood fell down upon the white drift below.
> "I wish my little daughter might have skin as white as snow, lips as red as blood, and hair as black as this ebony frame," said the queen thoughtfully.

Some correlations become apparent as we see evidence of our knowledge of biology at work in the conception process:

red	white	black
blood	snow	ebony
blood	semen	womb(?)

A second reference to the red/white theme occurs in one version where the queen salts, cooks, and eats the "heart of Snow White." This follows an old belief that eating the organs of animals (or people) will imbue one with their powers.

As the wicked queen prepares the apple that will finally succeed in eliminating Snow White from the picture, we see again the color symbolism since only half the apple is poison:

red	white
poison	pure
death	life

After Snow White's apparent death, the dwarfs perform a ritual of bathing her with water and wine. In response to the old belief that illness was caused by evil spirits, this ritual was a way to drive them out and restore health. If we compare the wine to blood and the water to semen we have again the symbols of conception, this time perhaps indicating rebirth. At the wedding the queen dances herself to death in red-hot iron shoes, carrying the color symbolism from beginning to end.

Van Gennep (1960) states that all rites of passage contain the three stages of separation, liminality, and aggregation. Snow White's separation phase occurs when the queen orders the huntsman to take her into the woods and kill her. The aggregation phase is, naturally, her rescue and marriage to the Prince. Snow White's period of liminality, living in the woods with the dwarfs, is marked by many of the properties listed by Turner (1974): absence of property, status, rank (Snow White no longer living in palace); humility, unselfishness, equality (keeping house for the dwarfs); anonymity (dwarfs know her identity but keep the secret); total obedience (expectation that she will heed their cautions); foolishness (her response to the "peddlar woman"—queen); silence (alone in house while dwarfs are working); sexual continence (this seems indicated while, interestingly, two versions state that instead of waking Snow White when they found her the seventh dwarf took turns sleeping one hour at a time with each of his comrades until morning; this could correlate with Turner's reference to bisexuality!). While the dwarfs are men, they take on the attributes of women, especially of a "godmother" as they caution and advise Snow White.

The comparisons to darkness and wilderness are represented by the forest; those to death and the womb by the coffin and house; and Snow White's acceptance of pain and suffering by the repeated threats to her life. Together these properties comprise some situations familiar to us in American society, from the monastery/nunnery type of life with its silence, humility, and poverty vows to the "dropouts," "hippies," "teeny-boppers," and commune groups who have been criticized for their attitudes toward work, sex, and "society."

Looking again at Snow White, this time with a Freudian (via Bettelheim 1975) magnifying glass, we discover some of the psychological perspectives relating to kinship:

Mother	*Father*
real=dead	real=good
step=bad	huntsman=good
dwarfs=good	prince=good

Viewed from this angle we can determine some of the fears in operation in both child and adult in this story of mother/daughter conflict; this makes the tale meaningful to individuals of any age. Fear of losing one's parents; fear that one might be "adopted" instead of the real child of one's parents; fear of growing old; fear of losing one's attractiveness—these are obvious. In our culture we hear frequently of a young-looking mother becoming competitive for her daughter's boyfriends, and then there is the usual struggle between mother and daughter for the father's attentions and affections. All of these point up the American emphasis on the importance of beauty, youth, and competition.

Since every myth should be examined also in the context of the culture of its origin, it is wise to consider it from the perspective of the period, looking at relevant historical events that may have spawned the themes. Snow White, like many fairy tales, came out of the feudal age where there was a distinct separation of elite and peasant segments of society. There was virtually no chance for upward mobility, although there was always a chance of losing one's birthright. Peasants lived by a fatalistic code with no real hope of bettering their position. Thus, fairy tales became a kind of release mechanism for the otherwise intolerable way of life. Some of the same attitudes toward life are prevalent in urban poverty areas, so there is a hint of this in *West Side Story*, but more relevant is the influx of Puerto Ricans into New York, which causes conflict as the various ethnic groups try to maintain their individual cultural differences and values in a country that almost compels them to conform to an ideal viewed as far superior to any "foreign" culture.

For this short paper, Wasylyshyn was asked to apply neo-Freudian analysis, as presented by Bruno Bettelheim (1975), to one manifestation of American "pop" culture. He chose a pulp literature series geared toward preadolescent boys. Wasylyshyn extends his analysis of the *Hardy Boys Mystery* series in the next selection, but here his concern is with the appeal of Frank and Joe Hardy to latent-period boys. Wasylyshyn demonstrates the Hardys' successful resolution of earlier stages of psychosexual development and the series' stylized splitting of positive and negative images of parents into separate good and bad figures.

9 / A Neo-Freudian Analysis of the *Hardy Boys Mystery* Series

John Wasylyshyn

This paper is a neo-Freudian analysis (as in Bettelheim 1975), of the *Hardy Boys Mystery* series books. On the back cover of each *Hardy Boys* book it is stated that the mystery series is intended for boys aged ten to fourteen. According to Freud, boys of this age would be classified as being in the latent stage of psychosexual development. Therefore, we would expect to find that the focus of the books is on the resolution of this stage, and it is. Frank and Joe Hardy are in their late teens, and are thus no longer in the latent stage of development. They have, in fact, successfully resolved the "crises" faced in each stage of psychosexual development. Thus, contained in the books we would expect to find evidence of this successful resolution of virtually all of the developmental stages, and we do. In addition to discussing the Hardys' resolution of the various crises of psychosexual development, and the books' emphasis on the latent stage, I shall also relate other aspects of myths and fairy tales to the *Hardy Boys* books, including Bettelheim's comments about idealized portrayals of "good" and "bad" parents. Also discussed will be how Freud's ideas of "id", "ego," and "superego" are symbolized in *Hardy Boys* books.

Briefly, Freud's theory of psychosexual development states that a person passes through a number of developmental stages from infancy to adulthood, and he or she must resolve a crisis in order to progress from one stage to the next. If a person has difficulty in resolving a crisis at a certain level, he or she becomes "fixated" at this level, and will exhibit signs of this fixation in later life.

The first two of Freud's stages of development are the oral and the anal stages. Neither Frank nor Joe appear to have fixated at either of these levels because they don't exhibit such traits as overeating, nail biting, or smoking (oral fixations); nor do they exhibit retention (stinginess) or release (overgenerosity) which are characteristic of persons with anal fixations.

The genital or phallic stage is the third of Freud's stages of development. It is during this stage that children first become aware of sex differences. It is also at this stage that children must resolve the Oedipal crisis, or the desire to have sexual relations with their parent of the opposite sex. Freud believed that children resolve this by identifying with, or imitating, their parent of the same sex, thus hoping to capture the affections of the parent with whom they want to have sex. It is quite clear that Frank and Joe Hardy—boy detectives—resolved their Oedipal complexes via identification with their father, who is a famous detective. However, although resolution of the Oedipal crisis through identification may have been the primary reason why Frank and Joe wanted to be detectives, their present maturity is evidenced by the fact that they often work *with* their father to solve a mystery, rather than competing against him as they would were they fixated at this level. As previously mentioned, it is also during the genital stage that sexual identity is established. The Hardy boys seem to have had no difficulties in this respect. Both are "all-American boys" who are star athletes. The potential problem of two young men spending virtually all of their time together, and thus raising questions concerning their sexual preference, is triply averted. First is the fact that they are brothers, which makes it okay for them to be so close; secondly, they are partners in a "business," which further legitimizes their time spent together; and thirdly, they each have a girlfriend.

Upon resolution of the Oedipal complex, children progress to the latent stage. The *Hardy Boys* books are intended for boys in this stage. Bettelheim states that children are attracted to fairy tales that contain symbolic representations of the developmental stage that they happen to be in. The fact that well over fifty million *Hardy Boys* books have been sold is a strong indication that the books are relevant to boys in the latent stage. According to Freudian theory, it is during the latent period that children temporarily give up sexual desires, and instead turn away from their families and begin friendships with age-mates. Since the leaving of the family and establishment of friendly ties with peers has not been completed by the readers of the mystery series, but has been completed by Frank and Joe, the books serve as a guide for the successful transition through this stage.

Frank and Joe have many friends. The most frequently mentioned are Chet Morton, Tony Prito, Biff Hooper, and Phil Cohen. Tony is of Italian descent, and Phil is Jewish. The aim here may be to encourage the readers to broaden their circle of friends. As mentioned above, the making of friends is not the only occurrence during the latent period. A second and equally important feature is the children spending more time away from their homes and families. In the vast majority of *Hardy Boys* books, Frank and Joe spend more of their days and nights away from home than they do at home. So it is away from home that they have most of their exciting adventures. The underlying message in this seems to be that all kinds of exciting adventures await a child outside of the home. The boys' willingness to travel may serve as an inspiration to their readers. For example, the fact that Frank and Joe may go to South America to solve a mystery may make it easier for a boy to decide to spend a week at summer camp. The books also combine traveling with making friends, because wherever the young detectives go, they always make at least one good friend. However, it must be noted that wherever the Hardys go, each has at least one member of his nuclear family with him—his brother. This may be to discourage a youngster from running away from home and losing touch with his family.

The final stage of psychosexual development is the adolescent stage. This period begins at puberty and is characterized by increasing interest in the opposite sex. The Hardys have made it through this stage, but their intended readers haven't. Therefore this stage is not emphasized in the books, so although Frank and Joe each have a girlfriend, the girls play a minor role in their lives.

We have now examined the psychosexual development of the Hardy boys, and the conclusion must be that they have developed into well-rounded persons, with no obvious fixations. According to Freudian theory, a person's or ego's actions are the result of an interplay between the id, or pleasure drive, and the superego (analogous to the more common term, "conscience"). If the id, which has no regard for morality and ethics, had its way, the world would contain only selfish, amoral, pleasure-seeking people. However, the individual's id is opposed by the superego, which is morality. Thus, Freud argues that an ego's actions are a compromise between the desires of the id and the guidance and restraints of the superego.

Is this dynamic opposition represented in the *Hardy Boys* books? The answer is yes. The interaction between id and superego is represented in a minor way in the Hardys themselves. Frank is always described as the more thoughtful and serious of the two, while Joe is described as impetuous. In other words, the Hardys didn't simply progress through all of their psychosexual developmental stages (where Freud said control of both the id and superego are learned) and come out perfectly. Somewhere along the line, Frank's superego became slightly dominant over his id, while Joe's id appears more influential than his superego in governing his behavior. Together, however, the boys make up an ideal ego, but what id and superego attempt to influence this ego? The id which attempts to influence the Hardys is their best friend Chet Morton. Chet is constantly eating, and all that he thinks about is food and leisure. His overeating is an indication that he failed to control his id at the very first stage of development. Chet is forever trying to convince the heroes to forget about whatever mystery they are working on, and to relax and stay out of danger. Chet is written into every one of the mystery books, which means that like an id, he goes everywhere with his ego. Freudian theory dictates that a well-rounded person (ego) is one who can channel the libidinal energy contained in his id toward more socially acceptable ends. Thus in each book is a sentence which tells the reader that although Chet is disposed toward relaxation, the Hardys know that they can count on him in a tight spot. In other words, like any well-formed ego, the Hardy boys are able to put their id to work for them when they need him.

The Hardy boys' superego is their Aunt Gertrude. She is never satisfied with anything they do, and is convinced that if they don't heed her advice and premonitions they'll either get into serious trouble, or else amount to no good. Nevertheless, the mysteries always end with Aunt Gertrude proud of her nephews; i.e., their superego is satisfied with the Hardys' behavior.

A final feature of Bettelheim's analyses of fairy tales is the splitting of the positive and negative aspects of real parents into separate, idealized characters who are "good" or "evil." Frank and Joe Hardy have a "good" mother and a "bad" mother, as well as "good" and "bad" fathers.

Laura Hardy is the biological mother of the Hardy boys. She is their "good" mother as well. She never scolds the boys, and spends her time at home quietly worrying about them. The boys' "bad" mother is their Aunt Gertrude Hardy. She is forever admonishing them and looks at the bad side of everything they do. Also, she never openly praises

Frank or Joe. However, a difference between *Hardy Boys* mysteries and fairy tales is the inclusion of a sentence in each book which tells us that Frank and Joe both know that their aunt actually loves them, and is even secretly proud of their achievements. The difference is that the mature Hardy boys realize that their "bad" mother isn't all bad, and is well-meaning underneath, while wicked witches (etc.) are always portrayed as all bad in fairy tales intended for youngsters.

Fenton Hardy is the boys' "good" father in the series. He is also their biological father. He taught them the basics of detective work and continues to advise and assist them in solving mysteries. He has provided well for them, giving them a car and a speedboat. And he often rescues them from dangerous situations.

The "bad" father of the Hardy boys is whichever criminal is behind the mystery they are trying to solve. He does such things as discouraging the boys from taking the case, and then intensifying his threats in an effort to get rid of them. As was the case with the boys' "bad" mother, here too the division between good and bad is not always a clear one. For instance, the Hardys usually start out with a number of suspects, each of whom has both good and bad traits. Also, the villain's use of henchmen to do some of his dirty work represents him as not all bad, because he only does part of the bad work. *Hardy Boys* books are in effect just the processes that the Hardy boys go through in searching out their bad father.

In doing this symbolic analysis of the *Hardy Boys Mystery* series from a neo-Freudian and Bettelheimian viewpoint, I examined the psychosexual development of the Hardy boys, and came up with the conclusion that though they were not perfectly developed individuals, they had no major problems during their development. In addition, it was pointed out that together they did form a perfect ego, able to control both their id (Chet Morton) and their superego (Aunt Gertrude), thus behaving in a socially acceptable (or better yet, socially desired) manner. It was also suggested that the books functioned as behavior guides for their readers. And finally, taking a cue from Bettelheim's analyses of fairy tales, I identified the "good" and "bad" parents of the boys. Here it was implied that the well-developed person recognizes that no one is all bad, and that individuals must be examined from several different angles or viewpoints before a judgment may be passed on them, much as Frank and Joe must investigate each suspect thoroughly before determining which is the criminal. Thus we see that the popularity of the mystery series is not an accident, but that the books are popular because they are relevant to the development of their readers.

In this second look at the *Hardy Boys Mystery* series, John Wasylyshyn compares versions of *The Hidden Harbor Mystery* written in 1935 and 1961. He finds especially obvious changes in the portrayal of regional differences, the strength of kinship and family ties, and racial stereotyping. The contrast in portrayal of sex roles is somewhat less apparent. Wasylyshyn links these changes in juvenile pulp literature to changes in American society. Among his important findings is the decline of regionalism with the growth of the mass media and American national culture; this is particularly marked in the absence of negative stereotyping of southerners in the 1961 book. Wasylyshyn also notes a more positive image of blacks in the newer version. Another significant change is decline in the strength of family bonds, as the 1961 Hardy boys are much more independent of their father than the Hardys of 1935, and as an interfamily feud is transformed to an intrafamily squabble. This detailed and specific analysis of change in popular fiction exposes the cumulative result of many changes in American values and attitudes over twenty-five years.

10 / **An Analysis of American Culture as Presented in Two *Hardy Boys* Books That Differ in Time**

John Wasylyshyn

The *Hardy Boys Mystery* series, begun in 1927 by Edward Stratemeyer (under the pseudonym Franklin W. Dixon) and later continued by his daughter, has been a major unifying factor for American males. The books, which have sold over fifty million copies, have helped to enculturate generations of readers who look (or have looked) upon Frank and Joe Hardy as role models. As one might surmise, the earlier books of the series have become somewhat outdated, and have had to be rewritten. In each of the "modern version" books is the sentence, "In this new story, based on the original of the same title, Mr. Dixon has incorporated the most up-to-date methods used by police and private detectives" (Dixon 1961). This statement, however, is not altogether true. The Hardy boys' basic investigative method of following and eavesdropping on suspects has not changed from the original to the newer versions of the books.

What have changed, on the other hand, are the stories themselves. This change is often quite drastic. For example, in the two editions of *The Hidden Harbor Mystery* discussed in this essay, virtually the entire story line has been changed. What follows is a detailed examination of these two versions of *The Hidden Harbor Mystery*. The original version was printed in 1935, and it was most recently rewritten in 1961. Intermediate printings are not listed in the newer version, so it is unknown just how many times the story has been revised. I will attempt to relate differences between the two versions to various changes, trends, and phenomena of American culture. Topics to be examined and discussed include regionalism, kinship and family ties, women in society, and race portrayal. My hypothesis, then, is that changes in American culture are represented in the changes made in *The Hidden Harbor Mystery* from 1935 to 1961.

It is necessary to determine who reads (read) *Hardy Boys* mysteries, and thus who is (was) subjected to their enculturation. It was mentioned earlier that Frank and Joe are role models for their readers. Thus we should be able to get an idea of their readership by studying them. Frank and Joe are white. The fact that they own a speedboat and car, and that their father (in the newer versions) owns a private plane, places them in the upper-middle to upper class. This, however, does not necessarily mean that their readers are of the upper classes, but only that they have the possibility of reaching these levels. In other words, the boat and plane serve as goals which the readers may hope to obtain.

Thus I would say that their readership is not of the lower class. Further evidence that the books are read primarily by boys who are not members of the working class is the fact that the books have been barred from many libraries for being "badly written and repetitious" (Zuckerman 1976). This means that the books must be purchased, and although a single book can be bought for under two dollars, there are now sixty books in the series. Though it is highly unlikely that everyone who buys one book will end up buying them all, it has been my experience that no one can read just one.

This phenomenon is not entirely due to the quality of the books themselves. Some pressure to buy more of the books is put on the reader by the publishers. For example, on the back cover of each book since about 1940 (when the newer-style covers came out) is a paragraph which begins with the sentence: "All boys from 10 to 14 who like lively adventure stories, packed with mystery and action, will want to read every one of the Hardy Boys stories listed here." Further, there follows a list of every single *Hardy Boys* mystery, preceded by the question: "How many of these books do you own?" (Dixon 1961). In addition to all of this, the spine of each book has its number (in the series) on it, and I can remember feeling a tremendous sense of incompleteness because I didn't have all of them. Surprisingly, this feeling grew stronger as I bought more and more of the books! As mentioned above, the age of a majority of the readers probably falls between ten and fourteen. A boy of this age would have to be classified as being at an "impressionable" age (i.e., the books can indeed have an effect on the enculturation of the readers).

In an effort to facilitate the readers' understanding of my discussion of the two *Hidden Harbor* mysteries, I feel it would be helpful if I included a short summary of each book.

In the 1935 version, Frank and Joe are on board a steamship which crashes and sinks in a storm. They are accused of stealing six thousand dollars from a southern gentleman (Samuel Blackstone), whose life Frank actually saved. In order to clear their names, they travel to Hidden Harbor, the man's home town, to find out if Ruel Rand, whose family is feuding with the Blackstones (and who was also on board the sinking ship) stole the money. In the end it turns out that the Blackstones' black servant, Luke Jones, stole the money, and was also responsible for keeping the feud going.

The 1961 version is somewhat different from its predecessor. In this edition, the Hardys are engaged by the editor of a southern newspaper, who is being sued for libel by Samuel Blackstone because someone unknown to the editor printed an article in his paper stating that the Blackstone family fortune had been built on smuggling and receiving stolen goods from pirates. Since the rumor had been printed, the boys were to prove it was true. There is also a feud between Rand and Blackstone, but it is over the rights to a pond which, depending on how one would interpret an old will, could belong to either of them. Rand wants the pond because he wants to excavate an ancient Indian village, and he wants to find the buried family fortune to finance the excavation. Blackstone, on the other hand, wants to keep Rand away from the pond (and the buried fortune) so that no one can prove the pirate rumor true. A third man, Henry Cutter, is pressuring both Rand and Blackstone to sell their rights to the pond so that he can build a fishing club (he says). It turns out that Cutter is behind all the mysterious goings-on because he knows that there is a fortune in lumber submerged in the pond and he wants it for himself. The Hardys bring him to justice and find the Blackstone fortune (which is worthless, because it is in Confederate currency), and they prove that the Blackstone fortune was indeed gained illegally. They also help to end the feud which Cutter had started up again.

Although the story line has changed greatly over the years, the Hardy boys themselves have remained relatively unchanged. They are two years older in the newer books than they were in 1935, and although this may have some significance (e.g., boys in 1935 matured faster), I believe they were just aged gradually to make them seem more realistic.

Another difference between the books is related to the increasing importance of recreation in our society. In the first book, the boys are "strictly business." They do go to a play, but it ties into the mystery they are trying to solve. In the 1961 version, however, recreation and pleasure play a much greater role. For example, the only reason the boys, rather than their father, took the case was that Mr. and Mrs. Hardy were going on a Caribbean cruise, during which Mr. Hardy was to combine business with pleasure. The Hardy boys themselves were actually reluctant to take the case because they had been planning a camping trip to Maine with their friend Chet Morton. In the end, they too decide to combine business and pleasure and go camping in Larchmont, Georgia, where the mystery takes place. There are also sequences in the book in which the boys go swimming or fishing and nothing related to the case happens. Further, the business and pleasure aspect is symbolically represented in the book when the Hardys have a chase scene through the fun house of an amusement park.

The modern-day Hardy boys are much more independent of their father than were the Hardy boys of 1935. As mentioned above, the boys take on and solve an entire case while their parents are on a cruise. Rather than asking permission, they simply send a telegram to their parents informing them that they have taken a case in Georgia. In contrast, the "original" Hardy boys rarely made a move without their father's advice. In fact they would never have been on the sinking steamship if they had not been running an errand for him. This difference in the boys' independence from their father may be related to the amount of ascribed and achieved status the boys have in each of the books. Although they have solved thirteen mysteries previous to this one, the boys are only recognized by strangers as "Fenton Hardy's sons" in the first book. This is a good example of ascribed status. Further evidence of their status being ascribed in 1935 is the author's statement that Frank and Joe had "inherited" their father's sleuthing ability (Dixon 1935).

By 1961, people had finally begun to hear of some of Frank and Joe's previous successes. When their eventual client, Bart Worth, finds that Mr. Hardy would be unable to take his case he says,

> I've read in the newspapers of Mr. Hardy's great successes—I've also read that his sons often help him out, and that they have solved some tough cases on their own. [Dixon 1961]

Thus we see that while the boys are still associated with their father, they have achieved some status of their own.

The Hidden Harbor Mystery is a good story in which to study regionalism as a phenomenon of American culture because the mystery takes place in the South, and the Hardys are from the North. In general, regionalism was much stronger in 1935 than in the 1961 version. In the older version, when the boys observe the license plate of a suspect's car, they note that it is from a "southern state," and the state is the same "southern state" from which Mr. Blackstone hails. Since the reader is not told which state, he is apt to generalize what goes on there to the entire South. In the newer version, the mystery takes place in the town of Larchmont, Georgia. Thus, whatever takes place there may be peculiar to the town, and not typical of the whole region. By naming a specific setting for the mystery, the author has made it more difficult for the reader to generalize.

If one were to make up a list of the negative things that come to mind when a northerner thinks of the southern United States, it would probably include the following: slavery (along with rich plantation owners living off the slaves), the Civil War, feuds, and lynching mobs. All of these stereotypical aspects of the South are included either directly or indirectly in the 1935 version of *The Hidden Harbor Mystery*. Samuel Blackstone and Ruel Rand, the only two characters who appear in both editions, live in mansions surrounding Hidden Harbor. Readers are informed of the fact that this harbor was used in the pre–Civil War days as an unloading spot for smuggled slaves. Since both families owned the harbor, the implication is that they ran the slave-smuggling operation. Furthermore, not once is it mentioned that either Blackstone or Rand has any kind of an occupation. The implication is that they are both still living off the money made in the slave trade. The fact that they don't have jobs may also be further evidence of the importance of ascribed status during the time of the earlier book. In other words, the two men didn't have to work because their ancestors were rich.

The Civil War is brought up in the earlier book when the Hardys go to see a play entitled *The Spy From the North*. This production, playing in the South, was about the heroic adventures of a northern spy during the war. It's difficult to believe that such a play would actually have been popular in the South. It seems more likely that the author included this bit just to remind the readers that the South lost the war, and to imply that the only heroes of the war were on the northern side.

In the first edition of *The Hidden Harbor Mystery* there is a feud between the Blackstones and the Rands. When the boys inquire as to what started the feud, they are told that no one knows, or that the families have always been on bad terms. The impression that one gets from this is that the families are feuding just for the sake of feuding. The Hardys are told, however, that about thirty years ago the feud intensified greatly when Ruel Rand's father John was lynched and Samuel Blackstone "may" have had something to do with it. Here again, no real reason is given for the lynching, implying that this too is just something that happens in the South. Later in the book, the Hardys themselves come within a few moments of being lynched by an angry mob of farmers, none of whom get in trouble for it. This implies that lynching was still considered to be okay in the South in 1935. What young reader would want to travel to the South after reading in the *Hardy Boys* mystery that he could get lynched at any time and for no reason at all?

Both Rand and Blackstone are written into the modern version of the mystery, but they are no longer idly rich. Again, this may have something to do with the increase in the importance of achieved status mentioned previously. In this book, Blackstone is a wealthy businessman who appears to have started from scratch after his family's fortune, made from piracy and smuggling (not slaves), was buried in the Civil War days. Ruel Rand is now a college professor, and not very wealthy at all. He derives his status from his profession. The only mention made of the Civil War (besides the discussion of when the fortune was buried) is when the readers are informed that there is a monument to the war in downtown Larchmont. The readers are not told, however, which side, if either, is glorified by it.

In the 1961 version of *The Hidden Harbor Mystery*, Rand and Blackstone are still feuding, but this time there is a definite reason for the feud—a dispute over the ownership of the pond, due to an ambiguous will. Also, there is no mention of lynching. Furthermore, even though it is called a feud, Rand and Blackstone were not actively carrying it on until Henry Cutter (the villain) started it up again. Even then, they try to settle it privately.

Whereas the slavery, lynching, and feuding contained in the first book served to set the South apart from the North and the rest of the country, in the modern version we have the establishment of some unifying factors. First of all, when the Hardys' client Bart Worth tells them about Larchmont, he mentions that it is on the Atlantic coast. Bayport, the home town of Frank and Joe, is also on the Atlantic coast. Although the first book makes reference to a coast and shoreline, these are never identified as being of the Atlantic. The change of Hidden Harbor's history from slave smuggling to piracy is a second unifying factor, since the occurrence of piracy in this country's history is much more widespread geographically than is the occurrence of slavery.

In 1961, college professor Ruel Rand would like to find the Blackstone fortune so that he can finance the excavation of an ancient Indian village. Native Americans existed in virtually every part of North America. In other words, they are not peculiar to the South. Besides this, the fact that an Indian village existed in Georgia reminds readers that the South had a history prior to the Civil War and slavery, as well as reminding them that the South has something of cultural value to offer—not just feuds and lynchings.

A final unifying factor may seem to be trivial at first, but it may actually be the most significant, because it serves to unify the people of the South with the book's readers. When Frank and Joe are locked on the balconies of a house and their friend Chet walks by below, they call down to him to go into the house and free them. Upon trying the door and finding it locked, Chet "almost automatically" checks under the door mat and finds a key to the house. This act directly links Chet, someone familiar to all of the readers, to the southerners, because he and they share a convention about where keys are kept.

Kottak (1978*a*) has discussed the deterioration of the American family as a trend in our culture, and a breakdown of kinship and family ties can indeed be seen to have occurred between the two printings of the mystery.

In the 1935 edition the Blackstone-Rand feud was between unrelated families. Both Blackstone and Rand had relatives. Samuel Blackstone had a daughter Alice, who, although she had not seen him in years, rushed to his side and stayed there after learning that he was ill. Ruel Rand had a mother and a brother, and a father who had been lynched. The strength of the Rand family bonds is revealed in their reaction to the lynching of John Rand. Ruel's brother Ewald was driven insane because of it, and Ruel himself claimed to have been mentally affected by it. Furthermore, although Ewald was insane, Ruel and his mother still took care of him rather than committing him to an institution. Once, as the boys were watching the Rand mansion, they witnessed Ruel walking along the shore with Ewald for half an hour.

By 1961 the interfamily feud has become an intrafamily dispute. Blackstone and Rand have a common ancestor in Clement Blackstone, and his will split the family. In other words, whereas John Rand's death united his family, Clement Blackstone's death divided his family.

In the new version, Samuel Blackstone's main fear is that someone will discover that the family fortune was gained illegally, thus discrediting his family. This might lead one to think that "the family" is still important in the more recent book. However, I believe that Samuel is only interested in saving his own face, because Ruel Rand is equally related to Clement Blackstone, and that Samuel doesn't really care if his ancestor is disgraced. The difference is that Rand doesn't have the same last name as Blackstone. So, although it initially appears that Samuel is concerned about the reputation of his entire family, we see that his concern can just as easily be interpreted as selfish.

Whereas both Rand and Blackstone had living relatives in the 1935 book, the only relative mentioned in the 1961 mystery is Clement Blackstone. This is an indication that the author of the newer version did not feel at all compelled to include other family members. Thus, the importance of the family seems to have been less at the time of the last printing.

Further proof of the decline of the family in the more recent book is the fact that when Samuel Blackstone wants to prevent the Hardys from questioning a key witness (Jenny Shringle), he pays her to go visit some relatives in a town fifty miles away. The implication here seems to be that she had to be bribed just to get her to go visit her relatives. A second conclusion that can be drawn from this is that Jenny's relatives wouldn't have put her up if she couldn't have paid for her keep.

It is slightly more difficult to analyze the roles of women in society by comparing the two versions of *The Hidden Harbor Mystery*, because of the five women who appear in the two stories, two of them (Mrs. Hardy and Aunt Gertrude) virtually cancel themselves out (i.e., each appears to have changed relatively little over time); and a third (Blackstone's daughter Alice) only appears in the earlier book. Nevertheless, we can make some statements on the topic.

First of all, much more time is spent discussing Aunt Gertrude's culinary abilities in the earlier mystery. This could be because this was considered more important in times past. However, an equally likely explanation is that Aunt Gertrude has a much larger part in the older than in the newer version. But then again, the fact that when trimming her role the author chose to leave out the talk about cooking may indeed be significant.

Both books have a character who is a spinster. In the previous book she is Miss Pennyweather, and in the updated version she is Jenny Shringle. Miss Pennyweather informs Frank and Joe that at one time she was engaged to Samuel Blackstone, and also that John Rand asked her to marry him. In the former case, he chose to marry another girl, and in the latter she turned him down. Jenny Shringle offers no explanations as to why she never married. In the first book the author felt it necessary to include excuses, while the author of the second book felt none were needed. This discrepancy could reflect a slightly greater emphasis on a woman's need to marry in earlier times, but it isn't definite.

Maxine Margolis (1977), in an analysis of historic relationships between attitudes toward women and female extradomestic employment, comments on the very low percentage of working women during the 1930s. Working women were thought of as "thieving parasites" because they were taking jobs away from men. Alice Fox, who was Samuel Blackstone's daughter in the 1935 book, was a working woman. However, she wasn't taking a job from a man because her profession was acting. Furthermore, her mothering instinct proved stronger than her devotion to acting, as she left the play she was starring in just before the final scene, when she found out that her father was ill. Also, when Mrs. Rand was injured later in the story, Alice nursed her, even though her family was feuding with the Rands. The messages are clear: a woman should follow her "natural instincts"; and there is plenty for a woman to do around the home. Also, we can say that the "woman driver" slur existed as early as 1935, because Alice is involved in a car accident when she rounds a curve too fast, and in the wrong lane, and runs into another automobile.

Certainly the biggest change in *The Hidden Harbor Mystery* from 1935 to the present-day version is in the attitudes toward, and portrayal of, blacks. In the earlier book, the person behind just about every evil deed is Samuel Blackstone's servant Luke Jones. This is how he is described the first time the boys see him, aboard a train:

> At this moment a burly, thick-set negro, very loudly dressed and with a swaggering, arrogant manner, came along the aisle. He sat

> down, and propped his feet up on the unoccupied seat opposite. [Dixon 1935:83]

When a conductor comes through and "quietly" informs him that it is against the rules to put his shoes on the seat, the following takes place:

> "Fool rules!" grunted the negro disdainfully. Nevertheless he removed his feet, but as soon as the conductor had disappeared, he put them back onto the seat again.
>
> "Luke Jones don't stand for no nonsense from White folks!" he said audibly. "Ah pays mah fare, an' Ah puts mah shoes where Ah please." [Dixon 1935:84]

The racial stereotypes of loud dress, arrogance, and insolence are obvious. What is less apparent is: first, that the seat next to Luke is empty, even though it is mentioned later that the car is very crowded; secondly, he is given quite an accent by the author, even though persons such as Samuel Blackstone, who is described as having spent all his life in the South, are given no southern accent. Since the way in which a person speaks greatly influences the opinions of the speaker formed by listeners, I am quite sure that readers of Luke's dialogue would not form a very high opinion of him.

The second time that the Hardys encounter Luke, the author writes:

> The more he saw of Luke Jones the more he disliked the fellow, who was dressed in a suit of extreme collegiate cut, and wore a pink shirt with a violet necktie. A diamond ring twinkled on his finger, and his patent leather shoes shone. [Dixon 1935:140]

The point to notice here is that this is only the second time that Frank has ever seen the "fellow" and still has no idea that Luke Jones is the villain, yet he already dislikes him. Furthermore, the diamond ring mentioned was stolen by Jones from Blackstone. Obviously Blackstone must have had it "twinkling on his finger" at one time or another, and I doubt that it was cause for dislike. The implication seems to be that blacks shouldn't be allowed to have expensive things. Finally, still another conclusion which can be drawn from the meeting is that Luke Jones is very stupid because he is wearing a stolen ring around the very household from which he stole it.

The Hardy boys go on to discover that Luke is the leader of a "secret society" of black youths (the implication is that blacks have to hang out in gangs). This group goes on to kidnap Ewald, the insane brother of Ruel Rand, and seriously injure his mother in the process (they pick on women and disabled people). Somehow Ewald manages to escape and take their guns at the same time (even a half-wit can outsmart them). Ewald goes on a rampage which incites a lynch mob to go after him. But we're told that it is not the mob's fault, because they don't know it is Ewald they are after; and it isn't Ewald's fault, because he isn't responsible for his actions. Therefore, the near-lynching of Ewald, and of the Hardys who are trying to aid him, is the fault of the blacks. From this, one might conclude that the previously unexplained lynching of John Rand may have been the fault of some blacks; and that perhaps all southern lynchings were really the blacks' fault. As farfetched as this may seem, if my mind could make these connections so could the subconscious mind of a young reader.

After Luke Jones confesses to all of his crimes, the author writes:

> Luke Jones, you're the worst scoundrel that we've ever run across," exclaimed Frank, "and we have run into some pretty bad customers during all of our experiences."
>
> Instead of being abashed by this statement Luke seemed actually flattered by the distinction. [Dixon 1935:200–201]

Not only is this role model (Frank) condemning the black man, but he is at the same time confirming that he (Frank) was correct in disliking him from the start. Also the black man (all blacks) likes the idea of being a scoundrel. Remember, many 1930s readers may never have seen a black person (Louise Berndt, 1979, personal communication).

As one might guess, the 1961 version of *The Hidden Harbor Mystery* does not include Luke Jones. Instead, he is replaced by two characters. The first is Timmy, a young black boy who is the grandson of one of Blackstone's servants. Timmy is a very nice boy who actually helps the Hardys solve the mystery. He has no trace of an accent, and even has a vocabulary that suggests that he is intelligent. However, no mention is made of Timmy's parents.

In describing the villain, Henry Cutter, the author seems to be trying to make up for the derogatory portrayal of Luke Jones, and all blacks, in the earlier version, by making Cutter as nonblack (white) as possible. Cutter is described as "a tall, *pale* man with *blond* hair, wearing a *linen* suit, and *straw* hat" (emphasis added).

It is unfortunate that the misconceptions and unflattering views presented in the original *The Hidden Harbor Mystery* appeared when the books had probably their greatest enculturating power. In 1935, the books did not face the competition that they do today from television and other mass enculturators such as popular music. Also, the book's enculturating influence is even greater when one considers that it was probably at least five, and probably closer to ten, years before the book was first rewritten; and even greater still when one takes into account that the book continues to enculturate even today, as the original readers of the book raise their children.

I believe that *Hardy Boys* books clearly exemplify the period of American culture during which they were written, not because the authors have consciously tried to write them with this goal in mind, but because they simply have attempted to make their books relevant, believable, and saleable. Similarly, I don't feel that when rewriting the book, the author consciously had Chet "automatically" find the key under the mat because he sought to create a unifying factor, but merely because the solution to the problem was plausible to him, and it seemed so because he is a member of the culture about which he was writing. That this solution may not have seemed reasonable to an author writing in 1935 merely stresses the notion that one is a member of a culture in both time and space.

Research Projects: Observation of Public Behavior, Researchers' Study of Their Own Groups, Interviewing, and Hypothesis Testing

The essays in Part 3 illustrate still other analytic techniques and subject matter. In contrast to the selections in Parts 2 and 4, none is an analysis of the mass media, and most do not employ structural, symbolic, or neo-Freudian analysis, although many exemplify an etic research strategy. The first three essays in Part 3 are by professionals. In essays 11 and 12 Robins and Eckert describe findings of their ongoing research projects in American high schools. In essay 13, Meltzer looks at general American customs about humor—specifically, puns and audiences' reactions to them. Essays 14 through 20 are all student research project papers. Like Eckert, Robins, and Meltzer, Schlesinger (essay 14) observed behavior in a public place. Essays 15 through 18 blend observation of public behavior with study of groups to which the authors are linked by employment, kinship, or student status. Essay 20, a study of a high school swim team, is the only symbolic analysis, in this case not of the mass media but of a group to which the author belonged. Essay 19, based on formal interviewing with strangers, is particularly important, since it illustrates the value of well-designed hypotheses, adequate testing procedures—and the possibility of negative results. Through their careful research, Van de Graaf and Chinni discovered that a hypothesis that seemed intuitively likely did not hold up to testing. Note, however, that like the studies in Part 2, those in Part 3 attempted to minimize the inconvenience to the public of the course research assignment.

Lynne S. Robins, a graduate student in anthropology and sociolinguistics at the University of Michigan and a teaching assistant in Anthropology 412 in the fall of 1979, analyzes some of the linguistic and nonverbal behavior she has observed during her study of a Detroit area high school. She identifies three social categories in the high school, "jocks," "freaks," and the people in between. On the basis of eleven tape-recorded interviews she notes and explains differences in the way members of each category use these terms. She then shows how tension between freaks, who are at the bottom of the high school social hierarchy, and jocks, who are at the top, is expressed in territorial behavior. Rituals of rebellion occur when freaks invade the "pit," an area usually reserved for jocks, on the jocks' status-confirming occasions—football Fridays and homecoming week. This careful analysis draws on sociolinguistics (the study of the social components of linguistic variation) and ethnoscience (the study of classification systems and semantic contrasts). Robins's ongoing study has used formal and informal interviewing, participant observation, and observation of public behavior.

11 / High School Peer Group Classification Systems

Lynne S. Robins

In this essay, ethnoscientific and sociolinguistic methods are combined in order to analyze the social meaning of classificatory terms used in a high school context.

An exploratory study of the social classification system used by Michigan High School students[1] indicates that they categorize their peers on the basis of seven different dimensions of contrast (see table 1). The primary terms *freak* and *jock* name two categories defined by the aggregations of semantic components noted in table 2. A third category describes students who are neither freaks nor jocks (see table 2). The names of these categories also function as labels, and I will argue that these labels are purposefully applied to individuals and groups in order to maintain a ranked social order within the school.

The analysis that follows is based upon texts of eleven tape-recorded interviews, each of which was designed to elicit information about high school peer groups and the social needs they satisfy. Information about informants' backgrounds and interests was also collected.

1. The name of the high school described in this paper has been changed to ensure the privacy of participants in the study. Likewise, the names of all informants have been altered.

TABLE 1 Dimensions of Contrast for Category/Label Domain

1. Hangouts	
1.0	Hang out in school
1.1	Hang out in upper cafeteria
1.2	Hang out in middle cafeteria
1.3	Hang out in lower cafeteria
1.4	Regularly hang out in the pit
1.5	Occasionally hang out in the pit
1.6	Hang out at the library
1.7	Hang out in the halls
1.8	Hang out outside of school together
1.9	Hang out at McDonald's
1.10	Hang out at work together
2. Nature of Group	
2.0	Uni-seasonal sport group
2.1	Multi-seasonal sport group
2.2	School activity group
2.3	Enduring social group
2.4	Uni-seasonal social group
3. Group Composition	
3.0	Females only
3.1	Males only
3.2	Males and females
3.3	Freshmen only
3.4	Sophomores only
3.5	Juniors only
3.6	Seniors only
3.7	Freshmen and sophomores
3.8	Juniors and seniors
3.9	Jocks
3.10	Freaks
3.11	Middle people
3.12	Band fags
3.13	Cheerleaders
3.14	Students from Central
4. Academics	
4.0	Go to class
4.1	Don't go to class
4.2	Do well in school
4.3	Don't do well in school
4.4	Some do well in school, some don't do well in school
4.5	College-bound

TABLE 1—*Continued*

5. Recognition	
5.0	Achieved status
5.1	Ascribed status
5.2	Labeled
5.3	Unlabeled
6. Degrees of Integration	
6.0	Wear recognized school uniform
6.1	Don't wear recognized school uniform
6.2	In student council
6.3	Time marker
6.4	Exclusion Marker (+)
6.5	Exclusion Marker (−)
7. Apparel	
7.0	Wear jeans with holes
7.1	Wear leather jackets
7.2	Wear down jackets
7.3	Wear makeup
7.4	Don't wear jeans with holes
7.5	Don't wear makeup
7.6	Wear high heels
7.7	Don't wear high heels
7.8	Wear jackets with school emblem

With the exception of one group conversation, all interviews were conducted on an individual basis.

The eleven students interviewed are members of a single social network. According to an outside informant, Bobby, Gayle, Michelle, Kathy, and Don are jocks; Joe and Ron are freaks; and Janet, Joyce, Rick, and Kevin are neither freaks nor jocks. Particularly striking is that my own sorting of individuals into clusters on the basis of interaction patterns coincides nicely with my informants' categorizations.

Interviews confirm that these categorizations would be generally acceptable to the members of this peer group. Each of the individuals classified meets the objective criteria of category inclusion specified by peer group members themselves, and noted in table 2. However, individuals in this study sample would not necessarily label themselves as they were labeled by my informant. The distinction I am making here is not trivial. The process of categorization is a linguistically neutral phenomenon which involves the grouping of individuals according to em-

TABLE 2 Composite Componential Definition of Categories (from table 1)

Jocks	In-between People	Freaks
1.1,2,4,7,9,10	1.1,2,7	1.3,9
2.0,1,3		2.3
3.1,2,3,8,9,10,13,14		3.1,2
4.0,4,5		4.1,3
5.0,2	5.3	5.1,2
7.3,4,6,8		7.0,1,2

pirically specifiable traits. The actual process of labeling, or the strategic assignment of a category name to an individual or a group of individuals, however, is somewhat different. Implicit in the act of labeling individuals is an evaluation of their status in the high school social order.

Michigan High School possesses a ranked social order that I will treat as a given. A follow-up study will be conducted to determine the socioeconomic and ideological correlates of its formation. But for the purposes of this essay, it is enough to know that, within the school, freaks are subordinate to people who are neither freaks nor jocks. And jocks are superordinate to both of the other groups.

Those most assimilated into the school's social system achieve the highest status. Social status accrues to individuals who actively participate in school organizations, attain elective offices, and compete for, attain, and excel in athletic positions. Academic achievements do not appear to play a large part in determining status, though both cutting classes and doing poorly in school are negatively valued.

That labels may be used to stigmatize is evidenced in the mitigation and avoidance techniques that they motivate. The following statement, made by a cheerleader named Gayle, is illuminating in this regard.

> We have one girl on our squad who is, well, what they, you know, she's like a freak on our squad. That's what they say anyway, that's not, I don't like to say that but she's, you could, well, it's hard, like you could pick her out of our squad.

Gayle admits that the girl described is identifiably different from the other cheerleaders, but she is reticent to label her a freak because of the stigma the label carries. Cheerleaders use the term freak to name a category of people who, in contrast to jocks, neither attend classes nor do well in school. Thus, implicit in its application is a negative evaluation of the person to whom it is applied. Not all labels are stigmatized, however. At Michigan High, the label *cheerleader* is a prestige marker.

According to Gayle, cheerleaders, like football players and basketball players, are distinctive by definition because their achieved status is socially recognized, and the labels accorded them translate directly into even higher status valuations. This is true not only in the general social order, but within a separate jock hierarchy as well. What separates cheerleaders, football players, and basketball players from all other jocks is that while all jocks have achieved status, only the individuals on teams that are labeled have the means to verbally display and enhance their status. To say "I am a volleyball player," is not socially significant, while to say "I am a football player" is.

Both Gayle and Kathy use the term *cheerleader* to flaunt their social ranks. In addition, nineteen of the fifty times that Gayle used labels and eleven of the forty times that Kathy used them represent instances of their using the label *cheerleader* to distinguish themselves from an unlabeled mass of jocks and regular kids and a stigmatized class of freaks. The amount of work that these girls do verbally to create and maintain their distinction can be measured. While Gayle and Kathy use labels fifty and forty times respectively, Janet and Joyce use them only nineteen and twenty-one times. Clearly, differences in labeling strategies correlate with social position.

Rick, Janet, Joyce, and Kevin are "regular kids," who occupy the middle stratum of the social hierarchy. Janet is the only member of this group who does not work, and her labeling behavior evidently diverges from the others'. Her social universe consists of three categories of people, the jocks, the freaks, and the nobodys, or as she says,

> not the nobodys, but the people who aren't classified as freaks and jocks but who are in between. We don't have a name for them, but they're there. There's no name for me, but there's a bunch of us who hang around each other and we're not jocks and we're not freaks.

Janet labels herself as the jocks would, and attaches the same value to social recognition as the cheerleaders do. Only a single semantic component defines her categories. Jocks and freaks are semantically equivalent to one another, in that both categories are defined by a +label component. In contrast to them are the nobodys, or the people who do not have a name, the people defined by a single −label component.

Kevin and Joyce refer to themselves as *the in-between people*, and *the middle group*, respectively, indicating that, while they recognize their places within the social hierarchy, they do not necessarily accept the values of those in the upper stratum. Rick is unique in that he divides his social universe into two categories of people, a work group and an in-school group. By avoiding the primary terms *freak* and *jock* in his labeling strategy, Rick verbally collapses the social hierarchy by neglecting to distinguish among its members.

Joyce's categories are not comprised of status-related dimensions of contrast, but she does recognize and label both the freaks and the jocks, and so verbally exhibits a labeling competence similar to the cheerleaders'. Unlike them, and like Rick, however, she distinguishes in-school friends from outside friends.

Janet's labeling strategy is almost identical to the cheerleaders', and I will argue that it diverges from that of the other members of the middle group because, like the cheerleaders, and unlike the other middle group members, Janet does not work. All of the social work that she does is done in school. The others leave school early to do real work in a restaurant. Not as much of their time is invested in school, and not as much of their social selves are. They do not have to work as hard in school to create and maintain status distinctions that they know have no value outside of a limited sphere of interaction. Less motivated to do the kind of social work they need to do in school, then, they remain members of a middle status grouping.

The labeling strategies used by Ron and Joe, who occupy the lower stratum of the social hierarchy, illustrate further that labeling behaviors differ with social status.

The componential configurations of the categories that Ron uses to classify others indicate that his labels, *the kids that smoke*, and *the smarter type kids*, are equivalent to the primary terms *freaks* and *jocks* respectively. What is peculiar about Ron's labels is that they explicitly, rather than implicitly, evaluate those to whom they are attached. While no status-related semantic components define his terms, neither is linguistically neutral. The same holds true for Joe's term, *burnout*, which is semantically equivalent to the primary term *freak*. I would suggest that this is a form of labeling incompetence, symptomatic of the freaks' social incompetence. While not meaning to deride, these two freaks do, by applying terms that explicitly evaluate. This contrasts with the jocks' strategies of implicitly deriding those to whom they attach labels. *The kids that smoke*, *the smarter type kids*, and *the burnouts* are labels that function like Janet's *nobodys* to maintain the recognized order, but, unlike Janet's label, they are not defined by a component of evaluation.

In contrast to the jocks, who actively engage in creating and maintaining a three-tiered hierarchy, the freaks do not distinguish a middle status category of people. Such a category would only serve to diminish their social standing, though I would argue that, given their lack of participation in the school value system, this would be of little importance to them.

Freaks oppose themselves to jocks using spatial rather than social dimensions of contrast. And these semantic oppositions have behavioral correlates.

The most frequently occurring component of terminological contrast contains information about the territories in which particular categories of people hang out. All the male jocks and all the male freaks contrast their categories using this component. Only two informants do not include territorial components in their category definitions and they are both female. Territory is the most cognitively salient dimension of semantic contrast for males in Michigan High School, while for females it is less so. An explanation of why this is so requires an examination of the "on the ground" territorial behaviors of peer group members in the high school setting, as well as an introduction to the territories themselves.

Desirable hangouts are a limited resource and the competition for them is evidenced by the tense encounters that an invasion of space entails. Jocks are the most mobile members of the high school community, and their territorial domination of the school creates a visual display of the privileges accrued to them by virtue of their status. While freaks have access to a single hangout in school, jocks have access to at least four.

There are qualitative as well as quantitative differences in the areas in which freaks and jocks congregate. The Michigan High School commons consists of three levels, each of which is claimed by different categories of people. The uppermost level is significantly smaller and less accessible than the other two. The middle and lower levels are comparable in size, but doors in the lowest level provide easy access to the outside. The freaks hang out in the lower commons because they can take advantage of its easy access to the outside, where they can smoke. They do not, however, restrict that access to others.

The jocks claim the uppermost tier of the commons as their own because, given its location and its size, it is easy to sanction intrusion. This territory keeps the jocks spatially distinct from the regular kids and the freaks in much the same way that the labels they use keep them socially distinct.

The social order that the jocks have created and maintained through the use of labeling is reflected in the high school commons seating hierarchy. No real territorial conflict occurs here, however, because one's place in the hierarchy does not restrict one entirely from the privilege of eating and hanging out in the entire commons area.

Completely restricted areas, however, are focal centers of jock/freak tension, and ritual reversals of territorial, and, by implication, social order take place within them. One such area is the pit, designed, ideally, to be a social center available for use by all students. This large, carpeted space, however, is used exclusively by the jocks, who, via implicit and explicit intimidation displays, effectively restrict its access to all others. The jocks' territorial domination of this scarce and coveted resource is a visual reminder of their superiority and of the privileges that go with rank and social acceptability.

Invasions of the pit occur predictably. During homecoming week, and on Fridays, when the male jocks are dressed in their suits and ties, and the female cheerleader/jocks are dressed in their cheerleading uniforms, freak/jock tensions mount. When the jocks are most vividly displaying their status, the freaks challenge that status by reversing territorial ownership and, by extension, the social order that supports that ownership. When these challenges occur, it is the male freaks and the male jocks who clash most violently at the pit, perhaps explaining why the territorial component of contrast occurs most frequently in the categories of the males than the females in this peer group.

Both order, and social order, are restored when school officials step in and usher freaks out of the pit, thus ritually reaffirming the status hierarchy.

Frequent association with the pit makes anyone a jock, from the freaks' point of view. Thus, pom-pom girls, members of unlabeled sports teams, and other nonfreaks who are allowed to congregate at the edges of the pit become jocks simply by association with the pit. And this serves to explain why freaks do not distinguish middle-status people from jocks.

Implicit in ethnoscientific investigations of terminological systems is an assumption that the classificatory competences of their users are most accurately described by single paradigms constructed of culturally salient semantic components. Any variations detected in linguistic labeling strategies, then, must be thrown out, or treated as insignificant deviations from a norm. In this exploratory study of one high school peer group's system of social classification, it becomes readily apparent that some group members are using different terms to describe equivalent categories of people and that individuals use a variety of nonidentical componential sets to define the terms they use. When I was viewing this study as an ethnoscientific endeavor, it became problematical because there was no way to justify a dismissal of the variations observed. Yet it would have been necessary to ignore, or factor out of my analysis, any terms or components that deviated from the norm. Rather than take those measures, I put aside the assumptions I had made about the existence of a norm, and then, using techniques borrowed from sociolinguistics and ethnoscience, I turned my attention toward studying the actual linguistic behaviors of my informants. The results of this investigation suggest that variation within a system is neither random nor trivial. It is, rather, a salient component of social meaning, and, as such, it cannot, and should not, be ignored.

Eckert's findings about manifestations of socioeconomic stratification in the contemporary United States are based on a study that began in 1980 of another high school in suburban Detroit. Here Eckert correlates a feature of clothing—the contrast between straight-legged jeans, flares, and bell-bottoms—with social categories in the school population. Her study is both quantitative and etic, based on observations of the width of blue jean bottoms in the territories of "burnouts" and "jocks," and in intermediate social space. Eckert also notes that this social marker is observable in social pairings. Although we all know that socioeconomic class affects clothing styles, Eckert's study provides specific and detailed confirmation in one contemporary American setting.

12 / **Clothing and Geography in a Suburban High School**

Penelope Eckert

While the character of the population of every community differs across both time and space, the structural feature of stratification is constant. In the same way, although student populations of no two high schools are alike, these populations tend to be organized by constant structural features. The most striking of these features is the development and polarization of a number of individuals into categories that represent the two social extremes of the population. These two extremes have been found to correspond to the upper and lower ends of the socioeconomic scale within the local population (Hollingshead 1949; Larkin 1979), and they represent to the school population differences in relative assimilation to the values of the school. Students whose lives center around involvement in school activities tend to be from the upper end of the local socioeconomic continuum. They cooperate with and share many social values with the school staff. As a result, they enjoy considerable favor and freedom in the school and they generally feel that the school is serving their needs. Depending on the era and place, these students will be referred to by such terms as *jocks*, *in crowd*, *elite*, and *rah-rahs*. In opposition to these students are those who do not feel that the school serves their needs or cares about them, and who are generally alienated from school-associated activities. These students, typically from the lower end of the local socioeconomic continuum, will be labeled by such terms as *hoods*, *greasers*, *freaks*, and *burnouts*, again depending on their specific characteristics as determined by era and locality. These two categories, representing local extremes, are the main foci of attention in the school: virtually all students orient themselves to the behavior of the people in these two categories, and members of the two extreme categories define themselves to a great extent in relating to each other. Boundaries between the extremes are closely maintained, and salient characteristics of each group are terms in a series of oppositions: nonparticipation vs. participation in school activities; truancy vs. attendance; smoking vs. nonsmoking; and so forth.

In the school that provided the focus of this study, everyone recognizes the two major categories as the "jocks" and the "burnouts" (the actual local name used for the "burnouts" will not be used). The burnouts are described by all (including themselves) as the kids who smoke, who take drugs or at least smoke some marijuana, who hang out in the courtyard of the school, who are not interested in school, and who do not care a lot about clothing. The jocks are described as the kids who are either abstemious or prefer alcohol to drugs, who are active in school activities, who go to class, and who dress well. Although these two categories do not account for half the population of the school, their polarized behavior defines the extremes of behavior for the school. The large mass of the school population—called the *in-between kids* or the *regular kids*—while perceived by the two extremes as an indeterminate mass, comprise a wide variety of groups, many of which overlap with the jocks and the burnouts. However small the actual number of "full-fledged" jocks and burnouts in the school may be, their significance in the school far outreaches their numbers. The quality of relations between jocks and burnouts governs aspects of behavior of virtually everyone in the school, for those who are neither jocks nor burnouts are nonetheless aligned or are struggling not to be. Since the jocks and the burnouts represent the two extremes of behavior in the school (with relation to the school and its values), it is a reasonable starting point to assume that the rest of the school population orients its behavior to that of the two categories. In this sense, the "in-between kids" are truly in-between.

Certain kinds of information about categories of people can only be obtained through observation of larger patterns of activity—through the movements of large numbers of members of each category. To do this, one needs a way of identifying the category membership of anonymous individuals. In a situation of extreme polarization, as between jock and burnouts, the possibilities for observable physical differences between members of the two categories are great. As people are thrown into intensive contact in the small area of a school, boundary maintenance is increased though careful attention to overt behavior: burnouts will not involve themselves in sports or school activities, jocks will not smoke, burnouts will not eat in the cafeteria, jocks will not enter the courtyard. These behaviors, however, do not provide ever-present category markers, since they are limited in time or place. Personal appearance is the only form of marker that an individual displays at all times, and that can serve as a consistent means of identification. Therefore, physical adornments such as clothing and hairstyle make ideal category markers if such markers can be found.

The clearest feature of dress that correlates with category membership in this school is the cut of jeans. When asked to describe the two social categories, students will occasionally mention that burnouts wear wide bells, and jocks are occasionally identified as always buying the latest fashions. The ratio of width of the bottom to the middle of the jean leg has changed over the years as an important component of fashion. The width of jeans that one sees in the school spans the past ten years of fashion, from wide bells through flares and straight legs, to pegged legs. This decrease in ratio of bottom to middle of the leg corresponds to fashion time, and in the larger context of Euro-American fashion, the continuum from bells to pegs is a continuum of fashion—the pegs being "in" and the bells being "out" at the time when this research was being done. Following fashion requires considerable financial resources, and insofar as socioeconomic class is an important term of the opposition between burnouts and jocks, the salience of jean width in this social context is clear. However, it would be misleading to imply that money is the direct and only cause of this difference in high school fashions, or that the burnouts are consistently "out of fashion." Since leg width has become an important category marker, the difference in leg width is maintained beyond financial limitations, and burnouts will choose not to replace their old bell-bottoms with new straight-legged jeans. And while burnouts may frequently have fewer clothing alternatives to jeans than people at the wealthier end of the social continuum, their general dress is not so much "behind" that of the jocks as it is in a different system. (The networks of spread of fashion change into and within the school population are themselves of great importance to study.)

The following discussion will demonstrate the usefulness of jean width as a marker of category membership, by tracing some large movements through the observation of jean width. Virtually all high school students own jeans, and on a given day, jeans account for more than half the lower body apparel of the student population (on one day, 52 percent of the girls and 64 percent of the boys were wearing jeans out of a sample of 190 students). Therefore, while jean width will fail to account for about 42 percent of the students, the sample defined by the wearing of jeans is sufficiently representative for the current purposes. For this study, jeans will be coded according to the relative width of the bottom of the leg to the knee. A value will be assigned to each style:

wide bells	4
flares	3
straight legs	2
pegged legs (and baggies)	1

and for any collection of individuals, the average jean width will be an average of these width values for jeans worn by those individuals. All other kinds of lower body apparel will be ignored.

The first problem in this study is to establish that jean width correlates with other aspects of category membership. As Robins mentions in her essay (chap. 11), territory in the school is a prime category marker. Territories in the school under study are occupied primarily during lunch periods and to some extent before school. The burnouts' territory was assigned, in a sense, by the school administration, when they declared the courtyard in the middle of the building as the only area in the school where smoking would be permitted. Since many burnouts smoke cigarettes, they go to the courtyard between classes and during lunch. And since smoking is a burnout category marker, the close association between the burnout population, smoking, and the courtyard has made this territory very strongly marked. One cannot say that the jocks have a territory in this school; it is the burnouts who have the territory, and the jocks' territory is defined residually by being as far from the burnouts' territory as possible. Most jocks report never having set foot in the courtyard during their entire time in the school, and virtually none of them ever go there. One can expect the average width of jeans therefore to be higher in the courtyard than in other areas of the school. In fact, this correlation is found in the data. The center of the burnouts' territory—the courtyard and its entrance from the school—is inhabited at lunchtime almost exclusively by people wearing wide bells; while the furthest area from the courtyard—the area directly in front of the cafeteria—is inhabited primarily by people wearing straight legs (pegged pants are still rare in the school). This is shown in table 1, which gives the average jean width in each territory and in the area between. Area 4 in table 1 is the entrance to the courtyard, the center of burnout territory, and area 1, at the westernmost extreme of the space

TABLE 1 Average Jean Leg Width at Lunch Period in Equal-Sized Areas of the Cafeteria Hall

West				East
Area 1[a]	Area 2	Area 3	Area 4[b]	Area 5
2.6	2.9	3.6	3.7	3.5

a. jock territory
b. burnout territory

available during lunch, is the jock territory. The spatial transition between these two core areas shows a gradual transition between mostly straight-legged and mostly bell-bottomed jeans. Moving eastward from area 1, jean leg width increases until one reaches area 4, and continuing past the center of burnout territory, the average width falls off again (area 5). This transition is the result of two kinds of mixture: between the two territories there is some interaction between people in bells and straight legs, and there is also a greater proportion of flares. Thus the area is transitional not only in average width but in actual width, indicating that flares themselves may be a transitional or "in-between" marker.

Examination of groups of students in the halls at various times, furthermore, shows that there is relatively little interaction between people wearing straight legs and people wearing bells. The socially in-between status of flares shows up again in the relative frequency of co-occurrence of the three kinds of jeans. If jean width were random—i.e., if people wearing all kinds of jeans mixed freely—one would expect to find all kinds of jeans co-occurring in proportion to their overall frequency of occurrence. Since there are the same number of bells as flares in the sample, one would expect a random mix to show straight legs co-occurring with bells about as often as with flares. However, in a survey of people walking together in the school halls, straight legs occur twice as often with flares as with bells. And while there are 50 percent more straight legs than flares or bells, flares occur about equally with other flares, with bells, and with straight legs. This confirms what was observed in the lunchtime territory dispositions: that flares are a socially transitional marker between bells and straight legs.

Once it has been established that jean leg width correlates with category membership, one can use it in observing general behavior. During class time, there is a certain amount of student traffic in the halls. Freedom to walk the halls during this time is in some sense a privilege, since it requires the explicit or implicit permission of the

TABLE 2 Average Jean Width in the First Floor Halls

West	East	Central		
		North	Middle	South
2.3	2.7	2.7	2.8	2.7

school staff. It is important to know who is in the halls, therefore, in order to get an idea of freedom of geographic mobility within the school. Jean width, therefore, was recorded for 400 students walking in the halls during the same half-hour class periods during one week, in order to get an idea of freedom of geographic mobility within the school. First of all, it is significant that whereas the average jean width in the school as a whole is 2.8 (based on a sample of 711 students), the average width in the halls during class was 2.6. Thus it is apparent that jocks are circulating more freely in the halls during class time. Table 2 shows the average jean width of people circulating in each of the five halls that make up the first floor of the school. While the value hovers around the overall average of 2.6 in the east and central halls, it is considerably lower (2.3) in the west hall. This hall is the area of highest visibility in the school, for it is the front hall of the school, running between the front entrances. It is also the riskiest area for walking, for it is the hall that the administrative offices are situated on. Thus it appears that the jocks are not only freer to move in the halls in general, but whatever burnouts are in the halls prefer to stick to the relatively unmonitored back halls.

One additional correlation in these data ties together the relation between jean width, geographic mobility in the school, and financial resources. The data used in this report were collected during the week that followed spring vacation. During spring vacation, large numbers of students traditionally go to Florida, and the Florida tan is an important status symbol in the school. Of the people recorded walking in the halls, seventeen were unmistakably tan. All but one of them were recorded in the front (west) hall, and their average jean width was 1.9.

Of course an indexical feature like jean width will not provide insight into the social structure of the school. It will only provide the means to observe correlations with other aspects of behavior that will suggest fruitful areas of inquiry. When one is doing research in a familiar setting, as a high school is to most of us, there is considerable danger of overlooking important phenomena. The use of an indexical device in quantitative observations may direct our attention to areas that we would otherwise have ignored.

William Meltzer, a graduate student in anthropology at the University of Michigan and Anthropology 412 teaching assistant, writes here of his observations of Americans' reactions to puns. Reactions are typically negative, violating the American cultural rule that "if you can't say something nice, don't say anything at all." The groans and hisses that follow a pun, Meltzer argues, may reflect its injection of play and irrelevance into nonplay situations, a signal that the punster is not taking the speaker and context as seriously as is appropriate. Yet Meltzer also shows that groans and hisses don't always mean disapproval, since they also follow good puns, those that are appreciated by the audience. Thus the response (e.g., a hiss) may, like the pun itself, mean something other than what it says on the surface. What in other contexts is negative response is here the culturally proper way of accepting humor and of permitting normal interaction to resume.

13 / **Jest Deserts: Audience Reactions to Puns**

William J. Meltzer

Anthropological considerations of humor have been oriented toward social order and solidarity. Thus Radin (1957:54) has mentioned the social control aspects of ridicule, and Radcliffe-Brown (1952:106–107) has shown that joking can provide an acceptable avenue for expressing tension and that it can bridge social disjunctures. Others have mentioned the general expressive solidarity of laughing at the same thing. However, the everyday meaning and significance of people's responses to humor are rarely explored. To shed light on the more general topic of symbolic interaction, this essay describes and interprets reactions to punning by inhabitants of the American Midwest.

Americans consider the pun to be a low form of humor (Benet 1965:826; Farb 1977:281). A few quick examples may help make this clear. In a brief review of a recent collection of punned haiku, the mind of the author is called "slightly bent" (Anon. 1979:5). The subtitle of a popular guide to punning is "How to Lose Friends and Agonize People" (Gordon 1980). An archaeologist, writing on a project in the Tehuacan Valley of Mexico, asserts that neither jokes nor bad puns affected his team's enthusiasm (MacNeish 1967:10). This last example is most telling, since the term *bad* is reserved for puns. That is, as opposed to other sorts of joking, puns, as a category, are singled out for stigma. All puns are equated with bad puns.

In actual interactions, people's responses to puns appear to be consistent with this assessment, as the following episodes taken from my observations of university students and employees show. A group of friends were at a tavern sampling various kinds of imported beer. One remarked, "This beer sure is jumpy. I wonder why." Another replied, "It must still have the hops in it." The others present responded with groans, winces, and cries of "That stinks," "Oh no," and so on. Another pun occurred during a dinner party. One fellow in particular was the butt of some lively repartee. Another, regarding the first fellow's very pregnant wife, commented, "And Betsy just grins," whereupon another person added "and bears it." Again groaning and "Oh no" were the responses. In fact, in almost every observed or reported case, puns were reacted to largely with groans, hisses, or other common expressions of displeasure. As one of my informants pointed out, puns are sometimes called groaners.

Why is this the standard reaction to puns? First consider the nature of puns and how they may cause difficulties in interaction. Puns depend on the ambiguity of meaning of the sound of a word. In a homonym-filled language such as English there are many opportunities for a punster to interpret a sound in a way unintended by the speaker, or to use a word that has a "trapdoor," a word that suddenly suggests a different level or context of meaning from that of the normal flow of conversation. In the surprise resides humor. Such use of homonyms is, in the cases dealt with here, playful. It is designed for fun. And it is this playfulness itself that can cause problems. That is, since puns are often simply slipped into normal discourse, they may interject a play framework into nonplay situations or conversations. Such framework switching, especially without warning or negotiation, can be confusing and disconcerting to others. Puns may thus violate a basic convention of human symbolic interaction, namely the understanding that responses to an act will occur within the same general framework or definition of the situation—as jocular or serious—in which the original act took place (Emerson 1973:275).[1] That is, the punster, instead of staying within an earnest frame of interaction, jokes.

In a similar vein it can be argued that puns may violate another foundation of communicative interaction: that of relevance. This refers to actors' tacit assumption that persons engaged in a conversation will keep their contributions relevant (Myers 1977:179). What happens when these rules are broken can be seen in the following instances. In one

1. For elaborations on the concept of the definition of the situation see Hewitt 1976, McHugh 1968, Stebbins 1967, and Wolff 1964.

exchange (between myself and a "native informant") a lack of regard for the initial earnest framework is revealed. I asked my informant why she thought people groaned at puns and she replied, "because the people who hear them are grown-ups," at which point I found myself spontaneously wincing and emitting groans.

In another case a student was showing slides of excavations he had helped carry out. This was at an informal student gathering. Pointing to one slide he said, "This skeleton was found buried in that pile of shells." A member of the audience then spoke up, "Did it have clammy fingers?" Amid groans and hisses the presenter responded, "I'll ignore that." Both he and the audience apparently condemned the pun. It had been abruptly inserted into a situation that was basically nonplay, and it also posed an irrelevant, and hence irreverent, question to the main speaker.

By their nature puns will tend to violate both rules. As plays on words they project a play framework, and because they propose divergent meanings they often introduce an irrelevant interpretation or comment. Nevertheless, making intentionally irrelevant and playful remarks is hardly appreciated, since doing so implies a lack of regard for others' topics and seriousness. It is an affront to hearers' sensibilities. Or, in more academic terms, by forcing people to consider various possible meanings of uttered sounds, instead of largely assuming those meanings from the context of interaction, punsters disrupt the routine thematic organizations of meaning that persons use to guide their conduct. Punsters often break other people's tacit trust that conventions of relevance and seriousness will be followed. The typical reaction to puns—groans—can be viewed as a negative response to these violations. Playing with the meanings of sounds and so jeopardizing underlying assumptions of interaction, the punster victimizes others and threatens the very basis of interaction—shared expectations—and is censured for such play. The lesson is that one simply must not toy with these communicational understandings.

Initially this line of argument may seem sufficient, but certain problems place it in doubt. The first arises from historical considerations. If puns do take liberties with the basis of human communication they should be condemned in all times and places, yet prior to the late seventeenth century, puns were quite popular in Europe (Crosbie 1977:1; Preminger 1974:681). However, since that time puns have been continually given bad critical assessments, being compared to thievery (Carleton 1877:143), and generally seen as indicative of disregard for rules (Preminger 1974:681). Although the reasoning of such critics often parallels that just presented, the point here is that this heritage of disreputability may *itself* be a factor in current responses to puns. That is, if our cultural tradition dictates that puns be condemned—and the historical record seems to support this contention—explaining groaning reactions to puns by analyzing their situational characteristics may be superfluous. Some of my observations suggest this possibility by showing that even when explicitly framed as play, puns elicit groans.

For instance, in one example a friend said to another, "Listen to this joke: you can lead a horticulture but you can't make her think," to which the other responded by groaning and saying, "Oh, that's a good one." A second example is taken from a Tom Lehrer comedy concert recorded before a live college audience in the early 1960s (*An Evening Wasted with Tom Lehrer*). Lehrer referred to a friend who "invented gargling, which prior to that time had been practiced only furtively by a remote tribe in the Andes who passed the secret down from father to son as part of their oral tradition." On the record one can hear groans, boos, hisses, and some laughter from the audience.

In both of these cases the general definition of the situation was one of humor or play. Jokes were expected, but instead of laughter the major responses to the puns were groans and boos. The relevance rule tends to be precluded in such comic frameworks. Thus, unless we wish to follow Oliver Wendell Holmes's (1960:15) simplistic contention that puns are an outrage against vocabulary—a contention that contributes to the tradition of censuring puns—we must consider the possibility that groans, hisses, and so on are often merely the culturally appropriate response to puns and do not necessarily carry a condemnation. This view is strongly supported in the two examples just given, since in the first the pun earns praise, and in the second some laughter is mixed with the groaning. Even more support for this position can be found in the next example, in which even though the framework is wrenched from serious to play the groans give way to approval. This pun was heard at a birthday party. When one person asked where the practice of blowing out candles on birthday cakes originated, another quipped, "Maybe it's to remind you not to make light of your birthday," a suggestion met by groans, and again the assertion, "That's a good one."

It appears, then, that groans and other such expressions of displeasure may in fact *not* be expressions of displeasure at all. The latent meaning of such responses may not be the negative sanction that is manifest, but, rather, acclaim. Indeed, in many of the examples presented here, and in others I observed, the groans and grimaces that puns elicited may well have indicated something like "We got the pun and appreciate it." This interpretation of seemingly negative responses to puns as positive evaluations also permits sense to be made of the following quote taken from the preface to an anthropology reader: "We were fortunate in being able to call on . . . the services of Mrs. . . . Walker for an unending supply of bad puns and good typing" (Arens and Montague 1976:x). It would be most peculiar to thank Mrs. Walker for an unending supply of something truly bad, and it seems clear that the term *bad* is meant to be understood as its reverse. Similarly I contend that groaning responses to puns are often tacitly understood by both punster and so-called victim as praise. Even people who enjoy puns groan at them, as my own reaction in an earlier example showed. And it is often observed that the worst puns are the best (Bather 1977:274; Lamb 1977:279).

A final example may lend further support to this by illustrating a definitely negative reaction to punning. Here a student and a professor were conversing informally at a bar. The professor asked the student, "What are you up to this term?"; the student replied, "Oh, about five six," and the professor, glaring at the student, said, "I don't know if I should continue this conversation. That's really lousy." Here groaning, wincing, and other such responses were absent. I contend that this is because such responses would tend to legitimize the pun, would accept without censure the deliberate playful (mis)interpretation made by the student. Instead, an unequivocal condemnation was forthcoming.

A brief recap of my main points may be helpful. I have argued that punning violates one or more basic requirements of human symbolic interaction—specifically, tacit expectations about others' conduct. Such violations make puns vulnerable to censure. However, the historical aspects of this present the possibility that audience responses, though typically groans, are not always condemnatory but, rather, are mandated by tradition. These two views—the historical and the situational—need not be exclusive. It might be proposed, for instance, that not only the origin but the retention of the groaning tradition is related to the rule breaking that punsters engage in. However, this slights the finding that groans and the like are often expressions of acclaim. The exceeding complexity of determining whether a groan expresses displeasure or pleasure, even for natives, ought to be clear. It could turn out to be the case that it expresses some of each. The groaned response may, for instance, be a form of playful retaliation on the part of a punster's audience. While temporarily accepting and responding in a play frame, the audience sets the stage for a return to normal discourse, restoring balance to the situation by playfully chastising the punster. Regardless, it would be valuable to investigate how other sorts of plays on words, such as spoonerisms and malapropisms, are responded to in our culture, and also how puns are treated elsewhere. Be this as it may, I'd like to suggest that the concept of the definition of the situation may eventually help resolve this problem by focusing on actors' understandings of what they and others are doing in some interaction. Since elements of such definitions are sometimes threatened by puns, it seems to be a good place to look to ascertain conditions of both positive and negative responses. For instance, it seems likely that when the general definition of a situation is one of play, or one allowing a bit of frivolity, groans will be of pleasure. Another possibility is that variations in persons' situational self-conceptions—one part of individuals' definitions of situations—may also influence the intended meaning of their responses. Thus, persons who see themselves as quick-witted may be more likely to groan with appreciation than those who see themselves as slow-witted victims losing face to others' rapierlike wit.

Basically I am urging that attention be paid to those variables contained within the general concept of the definition of the situation. This is necessary not only to determine the evaluative content of responses to puns but also to establish what conditions tend to accompany both positive and negative evaluations.

This could have some larger significance, since to the extent that the groans elicited by puns are taken by actors to be negative evaluations, they present an exception or contradiction to the general American cultural understanding that, at least to a person's face, judgments of them will be favorable unless unavoidably otherwise. That is, they would violate the guideline that if you can't say something nice about someone, don't say anything.

Another point worthy of note is the fact that the surface meaning of groaned responses to puns may not be the actual meaning. This draws attention to the broader category of cases of people saying one thing but meaning (and being understood as meaning) another. Irony is the obvious example. Such instances, precisely because of their complexity, are especially interesting and demanding of exploration. They hold much promise for yielding insights into the nature and conditions of human interaction, particularly in terms of the manner in which people anticipate others' acts in ongoing situations.

In this analysis of behavior in a public place (the University of Michigan Intramural Sports Building and the Imperial Court Club), Kenneth Schlesinger groups reactions of racquetball players to the points they lose into eight categories. On the basis of his observations over a three-week period (twenty-five games), he identifies the most frequent reactions as facial disgust, swearing, self-condemnation behavior, physical assaults on the racquetball court, and visual checks of the observation deck. Schlesinger's analysis shows the reflection of general American values, e.g., preoccupation with self-image and concern about others' evaluations, in racquetball play. Also important is his point that sports are worthy of anthropological study not just because they are popular but because they exemplify rule-governed behavior in demarcated settings, thus limiting the range of individual variation.

14 / **Reactions of Racquetball Players to Lost Points**

Kenneth Schlesinger

It is of anthropological interest that many Americans allocate much of their leisure time to participation in athletic activities. Anthropological attention has been devoted to explaining the popularity of athletics; however, the relevance of anthropology to competitive sports transcends the attempt to determine their appeal. Competitive sports provide an opportunity to observe people under controlled circumstances, since many athletic activities are characterized by standardized spatial arrangements, explicit rules, and criteria for success that are common to all who participate. The controlled nature of competitive athletics permits one to notice behavioral consistencies that might not otherwise be discernible, due to subjective differences in motivation or physical conditions.

For the present study I have observed the behavior of individuals engaged in competitive racquetball play. In particular, I have focused on perceptible reactions that follow loss of a point, and that precede the player's acknowledged readiness to resume play. I shall discuss the nature and frequencies of these reactions, as well as their meaning and significance to understanding American character.

I observed racquetball games at the University of Michigan Intramural Sports Building and the Imperial Court Club. Both of these athletic facilities are only open to people who have paid considerable sums for user privileges. The use of the Intramural Sports Building is contingent upon the payment of a substantial tuition fee; court privileges at the Imperial Court Club require a sizable membership fee in addition to court costs per hour. Since user privileges at both facilities are relatively expensive, it is possible that the population of athletes that I have observed is comprised largely of middle- and upper-class people. I observed both males and females (thirty and twenty, respectively) whose ages varied from the early teens to the mid-fifties. I believe that the sample population is sufficiently representative of middle- and upper-class Americans that valid generalizations about members of these social classes can be made. Although it is my conviction that the reactions of individuals to lost points do not vary significantly with respect to differences in social class, I have not attempted to substantiate this claim, and have qualified my results accordingly.

I observed racquetball games at these places over four weeks. During my preliminary observation (Week 1), I observed the range of losing behavior and noted which reactions occurred frequently. On the basis of my preliminary observation, I became conscious of similarities in the reactions of individuals to points they had lost. Although there were peculiarities in reactions, the observed behaviors could be classified into eight specific categories: (1) self-condemnation behavior, (2) physical assaults on the racquetball court, (3) stall tactics, (4) visual checks of the observation deck, (5) compliments directed toward one's opponent, (6) facial disgust, (7) swearing, and (8) indifference.

I observed twenty-five racquetball games over a three-week period, during which time I recorded the frequencies with which losing behaviors occurred, treating the occurrence of each reaction as a single observation of its parent category. Since all the reactions I observed were amenable to classification into the eight categories, I did not need to record any additional significant reactions, nor to determine their frequency.

Before presenting the frequency distribution of the observed losing behaviors, I shall clarify and exemplify the categories. Self-condemnation behavior refers to self-directed verbal remarks that express anger or dissatisfaction. Examples include: "I stink," "I can't hit the ball," and "I'm worthless." Physical assault on the racquetball court is self-explanatory, and encompasses reactions such as hitting the walls with one's racquet, stomping on the court, etc. Stall tactics refer to instances in which individuals delay behavior for more than fifteen seconds after the server has acknowledged readiness to initiate the next point. Observed stall tactics include facing the back wall instead of the front, pretending to tie one's shoes, and slowly drying one's grip.

Visual checks of the observation deck refer to gazes at the observation deck of their court. Compliments directed toward one's opponent involve praise toward the victor, e.g., "Nice shot," "That's tough," etc. Facial disgust refers to the expression of anger or dissatisfaction through frowning, sneering, or shaking one's head so as to convey disapproval. Indifference refers to the seeming absence of any reaction(s) to a lost point, i.e., nonchalance in resuming the ready-court position, without delay as previously defined. With the exception of indifference, all of the categorized behaviors may occur simultaneously; they are not mutually exclusive. For example, one can praise an opponent while frowning and hitting one's racquet against the wall.

The frequencies of the preceding categories are noted in table 1. Facial expressions of anger and disgust were the most frequent reactions. Facial expressions varied somewhat, but disgust was evident in each case. The categories of self-condemnation, swearing, and facial disgust had similar frequencies. Physical assaults on the court and visual checks of the observation deck were two-thirds as frequent as self-condemnation. Indifference was approximately one-third as frequent, and stall tactics and compliments about one-fourth as common as self-condemnation.

Although certain reactions were much more frequent than others, I shall conjecture as to the underlying meaning of all these categories of behavior. Such a discussion is clearly warranted because multiple observations of reactions of each category were made in almost every game.

The relatively high frequencies of facial disgust, self-condemnation, swearing, and physical assaults on the court suggest an almost unanimous distaste for losing points among the sampled population. All of these reactions express anger or displeasure about the event that preceded their occurrence. It appears that the people whom I observed were very

TABLE 1 Frequency Distribution of Classified Reactions to Lost Racquetball Points

Category of Reaction	Number of Observations
Facial disgust	210
Swearing	202
Self-condemnation behavior	184
Physical assaults on court	126
Visual checks of observation deck	120
Indifference	65
Compliments	46
Stall tactics	45

judgmental of their performance. They seemed to have continuously evaluated their performance on the basis of each point.

Self-condemnation, a very common behavior, also demonstrates players' preoccupation with self-image. Most of the self-condemnatory remarks claimed limitations in abilities for performance, such as, "I can't hit the ball," and "I can't win a damn point." Perhaps these individuals were broadcasting these remarks in an attempt to impose order on their immediate environment (cf. Assagioli 1973:51), for it appeared that they were expressing aspects of their self-images that made it increasingly probable that their performance, if unchanged, would correspond to their own and others' evaluations of their athletic skills. These remarks suggest a strong self-consciousness in the sampled population.

The widespread tendency to check the observation deck may be seen as a demonstration of concern over the publicity of one's performance. It is my impression that players looked up at the deck in order to determine whether anyone had seen them lose the preceding point. Furthermore, since self-image is in part a result of how (one thinks) others see one, these reactions may be a further demonstration of the player's preoccupation with self-image. For, in checking the observation deck, an individual may have been attempting to increase his/her awareness of other people's reactions to his/her unsuccessful efforts, thereby facilitating the modification or reinforcement of his/her self-image.

It is more difficult to explain the apparently irrational physical assaults that individuals inflicted on the walls and floors of their racquetball courts. It seems plausible that such aggressive behaviors might have been techniques through which players attempted to neutralize their responsibility for losing a point. Perhaps when a player slammed his/her racquet against the wall, he/she was transferring the blame for defeat to the racquet, which had allegedly "mis-hit" the ball, or the wall, which had been in his/her way. Similarly, when a disgusted player stomped his/her feet on the court surface, perhaps blame was being transferred to the floor, which had "interrupted" the flight of his/her last shot. I am suggesting that the observed individuals might have engaged in these aggressive behaviors in order to cleanse themselves of responsibility for their errors. According to this analysis, they were punishments of court and equipment to permit the defeated player to "save face" and maintain a positive self-image.

Compliments directed toward an opponent serve a similar function. If a player loses a point and then praises an opponent's performance, the loser's responsibility or blame for the loss is effectively neutralized, since a compliment insinuates that the outcome of the preceding point was beyond the loser's control; i.e., the opponent hit a strong, unreturnable shot.

Stall tactics were less frequent than the other reactions, but were still used in almost all of the games. They may be interpreted as strategies employed to improve one's chances for victory. The importance of these behaviors is that they manifest a strong desire to win. Since intentional delays between points did appear to perturb the opponent, it seems that in these instances the desire to win was so strong that illegitimate means to victory were employed.

Although I have labeled the eighth category *indifference*, absence of any noticeable reaction may not truly indicate a lack of concern or interest. One might be feigning indifference in an effort to appear invincible or unaffected by the opponent's efforts. However, indifference might also be a true manifestation of nonjudgmental awareness; here the title of this category is most appropriate. In the former case, apparent indifference is a strategy that demonstrates both a strong desire for victory, and a consciousness of one's self-image, which one manipulates so as to attain a psychological advantage. But in the case of nonjudgmental awareness, the desire to win and the degree of self-consciousness are very much reduced. Since it is not apparent whether indifference is a disguise or an actual state of mind, it is difficult to draw conclusions from this category, even though these reactions occurred fairly frequently.

Although all these reactions were observed within the context of competitive interactions on racquetball courts, the preceding behaviors are symbolic of various aspects of American character. I do not wish to succumb to the myth of classlessness in American society; however, it is my conviction that these reactions symbolize certain personality traits that transcend social class boundaries. It is on this premise that I discount any restrictions on the generalizability of the observed reactions which might emanate from the skewed nature of the sample population. Furthermore, since I have only suggested that the sample population might be disproportionately representative of the middle and upper classes, it is not necessarily the case that the composition of my sample is so biased that the following generalizations must be qualified.

I have claimed that certain observed reactions to lost points were manifestations of the player's strong desires to win. These desires were indirectly demonstrated by the displays of aggravation that followed lost points, as well as by illegitimate tactics that players instituted in order to improve their chances for victory. The frequently observed desire to win illustrates the premium that many Americans place on surpassing other individuals who participate in activities similar to their own. Similarly, distaste for losing that characterized the sample population typifies the frustration that many Americans experience as a result of unsuccessful bids for fame, recognition, and distinction among their peers. Therefore, these behaviors are symbolic of the strong consciousness and sensitivity of American people to their relative social position, which I assume to be at least partially contingent upon the outcome of various competitive interactions with others.

There is also significance to the previously established judgmental awareness of the sample population. Americans who have never played racquetball also tend to reflect upon and evaluate their behavior. This reflective, analytical state of awareness is a manifestation of the dominant, linear mode of consciousness in American society. Ornstein (1972:57) states that our culture is so thoroughly based on this active, linear mode that many Americans "have almost forgotten that other constructions of individual consciousness, other cultural styles, are even possible." However, anthropologist Dorothy Lee (1950) reports that the Trobriand Islanders have a total present-centeredness, in which all actions exist only in the present. Therefore, although the tendency to reflect upon and evaluate one's behavior as if it were a performance is characteristic of Americans, it is not a cultural universal.

Finally, the concerns relative to the publicity of unsuccessful behavior and the maintenance of self-integrity, which I have previously attributed to the sample population, are generalizable to the American public at large. Americans frequently attempt to conceal their mistakes in order to circumvent ridicule and public embarrassment. Similarly, Americans often assert that they are victims of circumstance, and they frequently pass blame onto others in order to "save face" and maintain favorable public standing. The significant element of self-consciousness implicit in the preceding generalizations should be emphasized, for a concern over self-image is prevalent among Americans.

The observation, interpretation, and generalization of the reaction of racquetball players to lost points demonstrates the amenability of competitive athletic activities to anthropological investigation. The ultimate value of this research not only derives from the significant contributions it makes to the current understanding of American character, but also from its illustration of the potential significance of certain behaviors that either elude our awareness or initially appear too irrelevant for anthropological concern.

Suzanne Faber took advantage of her job as a cocktail waitress in a local bar to explore relationships between income level, alcohol consumption, and tipping. Her data are based on participant observation, observation of behavior in a public place, and formal questioning of informants. Her most significant finding is a tendency for lower-income people to tip more than middle- and high-income people, a pattern that may reflect social insecurity or their identification with the bar employee. As in the selections by Larson and Van de Graaf and Chinni, Faber uses quantitative data to substantiate her conclusions.

15 / **Social Class, Tipping, and Alcohol Consumption in an Ann Arbor Cocktail Lounge**

Suzanne Faber

American culture is marked by diversity based on differential access to strategic resources. Various attempts have been made to prove that behavioral differences between representatives of the many social strata are observable. Speech patterns, style of dress, mannerisms, and so forth, have been analyzed as indicative of a particular background, a particular class. While class differences are indeed evident in many areas of behavior, questions nevertheless remain as to the specific circumstances in which differences may be discerned. Are class-based differences manifest in even the most commonplace, routine activities? The focus of my research is an attempt to answer precisely this question.

Through observation of individual behavior in a bar situation, based on examination of drinking and tipping habits, I have attempted to determine whether or not class-based behavioral differences in such routinized circumstances are apparent. Being employed as a waitress at a local bar has afforded me the opportunity to compile the necessary data.

Both the location and the atmosphere of this tavern proved conducive to comparative analysis of class. The fact that it is situated in a hotel implies that its patronage is not limited to local customers. It is located near an expressway off-ramp but still close to many large and small businesses. It is large, clean, and tastefully decorated. There is a dance floor and live entertainment, usually including the kind of pop and soft rock music that appeals to a variety of age groups. Compared to other bars in the area, it attracts customers from a fairly broad range of income groups, as will be demonstrated further on.

The data were compiled over four weeks, during which I worked ten shifts. I approached more than 150 people in order to recruit 100 participants. Due to the nature of my research, I did not wish to alter their normal behavior patterns. I explained that I was a student doing anthropological research involving income distribution of people who patronize local bars. They were unaware that I would be taking note of particular aspects of their behavior.

I needed to obtain three different types of information. Obviously I had to determine the individual informant's social class. Realizing that most people would respond to a direct inquiry about their social position by affirming their membership in the middle class, and recognizing the social unacceptability of asking such a question, I decided to use a less conspicuous approach. After showing them a card I had previously prepared, which listed five gross income groups, I asked them to write down on the back of their check the letter which corresponded to their own approximate gross income. The choices were as follows:

A. $0.00–$10,999
B. $11,999–$19,999
C. $20,000–$35,999
D. $36,000–$59,999
E. $60,000 and above

I chose to use such wide income margins primarily because I felt that if the list had been more specific or if I had asked the participants to write down their exact income, many would have felt the invasion of privacy too great and might have withdrawn their offer to participate. It is true that income level alone does not necessarily determine social class. However, since class is the sum total of a number of interrelated variables, including income, certain generalizations may be drawn. Usually someone who draws a relatively large income has better access to strategic resources and thus is in a higher class than someone who earns substantially less. Upper, middle, and lower classes, then, are not absolute concepts, but merely ways in which the sum of personal variables may be understood relatively. Accepting this as a premise, I divided the participants, on the basis of income, into the following categories:

A. Lower class
B. Lower middle class
C. Middle class
D. Upper middle class
E. Upper class

The second category of information needed for the study was alcohol consumption patterns. On the back of each participant's check I noted the number of drinks ordered and the frequency with which they were ordered. I discarded the information taken from participants who stayed for less than two hours. Classification of the participants with regard to consumption habits was as follows:

Four or more drinks per two hours = excessive
Three drinks per two hours = heavy
Two drinks per two hours = average
One drink or less per two hours = low

The third area of concern was tipping behavior. After the participants had left I made note on their check of the amount left as gratuity. The basis for classification here was as follows:

20 percent or more = excessive
15–20 percent = high
10–15 percent = average
10 percent or less = low

Upon completion of the data-gathering process I was able to correlate the results in an effort to determine whether or not class-based differences were indeed manifest in drinking or tipping behavior. The findings are documented in tables 1 and 2.

Examination of the data reveals some definite trends. Most striking are the class-based differences with regard to tipping. The pattern is that persons of the lower to lower middle class tipped more generously than individuals of the middle to upper classes. Excessive tipping is apparent among only 3 percent of the higher income groups, while among those with lesser means, about 30 percent left tips upward of 20 percent. (I should point out that one man who claimed to be in the bottom income category left a tip of $20.00 for a bill that totaled only $8.50.)

There are a few possible explanations for this. It may be that lower-class people, who are less confident about their social status than wealthier individuals, feel a need to overcompensate by tipping excessively. They may also feel a certain loyalty to their class and identify with their waitress, whom they see as another member of their own working class. Generous or excessive tipping may thus express a certain class "camaraderie." Conversely, high-income people, secure in their social and economic positions, are less likely to overcompensate. Most of them were cautious to leave a tip close to the traditional 15 percent.

TABLE 1 Alcohol Consumption by Social Class

Social Class	Drinks Consumed per Two Hours				Total People
	4	3	2	1	
Upper class	2	3	2	3	10
Upper middle class	4	8	9	7	28
Middle class	4	10	11	8	33
Lower middle class	4	6	8	5	23
Lower class	0	2	3	1	6
Total	14	29	33	24	100

Though it is difficult to pinpoint a specific cause for class-based differences with regard to tipping behavior, the trend is significant. The variation in the percentage of the bill left for the waitress implies that a number of other factors also dictate tipping behavior. These will be discussed further on.

The data fail to reveal any marked trend that might support the hypothesis that variations in alcohol consumption are class-based. Among all classes, between one-third and one-half drank at least three drinks per two hours, but most people consumed one drink per hour or less. It becomes clear that drinking is a cross-class phenomenon and that level of income has little to do with amount or frequency of consumption. (More significant, though not related to the current hypothesis, however, are the obvious implications of such enormous liquor consumption to American culture in general. The data support the notion that alcohol is the most widely abused drug in this country.)

The data seem to indicate that class-based differences are indeed manifest in such routinized behavior as tipping. Yet, after completion of

TABLE 2 Tipping Behavior by Social Class

Social Class	Percent Tip			
	20	15–20	10–15	10
Upper class	0	30	70	0
Upper middle class	4	18	71	7
Middle class	3	18	61	18
Lower middle class	26	39	22	13
Lower class	33	33	33	0
All social classes	10	25	54	11

the data-gathering process, I must express my own skepticism about the validity of its results. The conditions under which information was taken were not conducive to either my own nor my informants' objectivity. My appearance, sadly enough, was a factor. The uniform I must wear for work is a lowcut black leotard and a slinky slit skirt. It certainly did not enhance my credibility as an anthropologist. Moreover, my dual role as waitress and researcher often conflicted, as customers who felt they had to wait too long for service subsequently did not cooperate. In addition, although I tried to be inconspicuous in determining participants' class, the fact that at least some of them were interested in making a positive impression (with others at the table and myself, too) makes me hesitant to accept their claims. For example, one of the men who claimed to earn over $60,000 per year left his room key on the table with a note that said he would be interested in discussing the results of my research.

More generally, too many variables could have affected both tipping and consumption data for me to be certain of their validity. Frame of mind, and desire to impress a member of the opposite sex, or a superior, are some of the many variables that operate in bar situations. My own frame of mind must also be considered. Those people who seemed anxious to cooperate in my research were naturally treated better than those who were reluctant. Tipping behavior therefore was not accurately or objectively expressed. In addition, although I tried to carefully monitor the consumption patterns of the participants, my personal dislike of overt drunkenness, along with my responsibility as a waitress, might have caused me to avoid tables where I felt the customer had drunk too much. It is for these reasons that I feel that a repetition of the experiment using a different researcher might not result in similar findings.

Although I am skeptical about the validity of the data, I am certain of other (nonquantitative) trends that became evident during data gathering. I noticed significant behavioral differences along class lines in both my approach to the customers and their reaction to me. Many of the better dressed, more sophisticated individuals I approached took great offense to my efforts to recruit them. I received many threats that the manager would be told of my behavior, and one gentleman went so far as to remind me of my "place." The vast majority of the overtly less sophisticated customers I approached greeted me with a great deal of warmth and enthusiasm. Recognition of the social boundaries reinforced by class differences was thus at least tacitly expressed.

Another phenomenon also became clear in my research. Many of the participants expressed difficulty in accepting the fact that I am both a student and a waitress. A waitress is seen as a lower-middle-class worker, socially stagnant, while students are perceived as being upwardly mobile, the inheritors of high social positions. My dual role was thus somewhat incongruous in the minds of many, and in almost every case, the customers expressed their great surprise.

It becomes difficult to draw any conclusions, based on either my statistical or qualitative findings. In the first case, I doubt the accuracy of the data (for the reasons stated previously), and the degree of subjectivity involved in the second case would make conclusions suspect. Nevertheless, after having done the field research, I must affirm my belief that class differences do indeed exist, even in the most routinized, commonplace activities. Perhaps, given more objective and scientific circumstances, this hypothesis may yet be proven.

Self-Evaluation

I began the project with the best of intentions. I later realized that my approach and methodology were naive. I was not fully aware of the complexities involved in any field research project and I am sure that is evident in the paper. I did, however, learn a great deal. I have begun to be much more aware of subtle class-based behavioral differences among people in all situations, not just in the bar. Yet I am disappointed by the fact that I spent a great deal of time and suffered a lot of aggravation for what turned out to be inconclusive. Some of the things that happened were, in hindsight, somewhat humorous. Some of the "regulars" who found out that I was doing research now refer to me as "Professor," and leave larger tips in an effort to help me with my tuition. One gentleman offered to fly me to the Caribbean with him so that I could research the natives there; his wife tipped me twenty dollars after I told him off.

Like Suzanne Faber in the previous selection, Gail Magliano blended employment and anthropological research. Magliano found that Ann Arbor financial institutions express our cultural preference for right-handedness. Observing the architecture, design, and workspace layout of nine local financial institutions, she found that these businesses tend to place the customer service department to the right and the teller area to the left of the main entrance. Since customer service departments take in the most substantial deposits, while tellers mainly dispense money, banks, eager to increase their funds, assign the culturally preferred location to customer service. Magliano's paper is a good example of etic analysis, since none of the financial institution personnel were aware that the cultural preference for right-handedness was operating this way in their businesses. Thus, another expression of our tacit enculturation is revealed.

16 / **Right-Handedness among Ann Arbor Residents, as Expressed Particularly in Financial Institutions**

Gail Magliano

The purpose of this paper is to explore the preference for right-handedness by the residents of Ann Arbor, Michigan, and how this propensity toward right-handedness has manifested itself in local financial institutions (FI).

Information for this project was gathered through observations of behavior patterns; investigations of the commercial banks, savings and loans, and trust companies in the downtown Ann Arbor area; and research of literature furnished by FIs as well as various other related readings. The FIs covered in this project include the main offices of Huron Valley National Bank, Great Lakes Federal Savings and Loan, Ann Arbor Trust Company, and National Bank and Trust. Branch offices were also considered and included Michigan Savings and Loan, National Bank and Trust, Huron Valley National Bank, and Ann Arbor Bank and Trust.

I first became interested in the right-handedness of Ann Arbor residents while watching throngs of University of Michigan students rush along the campus sidewalks. These walks are somewhat unique in that they are generally wide and there is little along the way to distract pedestrians. For example, there are no store windows to be examined. The walks also accommodate an enormous amount of traffic during the period between classes when students rush from one building to another in order to attend their next class.

As each student strives to carry out her/his primary goal of getting from one place to another in the least amount of time, an interesting pattern emerges. There is an unofficial dividing of the walks so that a two-way traffic pattern is formed. Each student stays pretty much to her/his right-hand side of the walk and oncoming traffic passes on her/his left. Thus an orderly flow of traffic is accomplished.

Another form of traffic pattern which is right-handed is street traffic. All automobiles are driven on the right-hand side of the street. Whether these pedestrian and auto traffic patterns are related to each other or to some other influence is not clear, but the street pattern is a universal within the American culture of which Ann Arbor is a part.

Further evidence of the population's propensity toward right-handedness is found in their formal greeting and leave-taking. This is a rituallike action performed by two people. Using their right hands, the participants firmly grip each other's right hand. The joined hands are then moved in a prescribed up and down manner. This performance is always done with the right hand and, like the street traffic, is also a universal within the American culture.

A final example that demonstrates the residents' right-handed preference is the commercial enterprise that provides two doors side by side for customers to enter and exit. These doors are generally electronically controlled so that each swings only in one direction. In order to enter the store the customer is directed to pass through the right-hand door while the other door is used by exiting customers. The result is that both exiting and entering customers are allowed to use the right-hand door, i.e., right/in and right/out. Thus the proprietor succeeds in complying with her/his customer's right-handed preference both in coming and going.

When these same doors are not marked and do not direct the customer traffic, the people invariably go to the right-hand door, whether there is other traffic or none at all. Referring back to the campus sidewalk observations, this right-hand door phenomenon with no traffic would seem to indicate that right-handedness is a behavioral predilection for these residents.

Exceptions to the right/in and right/out door arrangement have existed in Ann Arbor. In the early 1970s, there was a grocery chain called Wrigley Food Stores in the Ann Arbor area which placed its doors backward, i.e., left/in and left/out. This arrangement caused countless shoppers, including myself, to walk into an automatic door which refused to open. A few years ago the Wrigley stores quietly closed. Perhaps the general lack of business which accounted for their closings was due to prospective customers constantly being upset when entering or leaving the premises and subsequently seeking out more comfortable establishments in which to shop.

If, as these examples indicate, Ann Arbor residents have a definite preference for right-handedness, how is left-handedness perceived? The noted anthropologist Claude Lévi-Strauss has argued that the human mind needs to impose order and that this need leads to classifying. One of the most common means of classifying is binary opposition, i.e., good and evil, old and young, big and small. It follows, therefore, that if right-handedness is preferred, then its opposite must be unpreferred. This assumption of negative feeling toward left-handedness is supported by the fact that no left-hand conventions were found among these residents and that their speech reflects the unpopular aspects of the word "left." Residents speak of being offended by a left-handed remark or, when not included in an event, they complain of being "left" out. In general the terms associated with "left" are rather negative, comprising such words as *sinister* (Latin for "left"), *awkward*, *radical*, and *clumsy*. As might be expected, the terms associated with "right" are positive and include such words as *justice*, *honor*, *virtue*, and *morality*.

Thus we see that the Ann Arbor residents have a definite propensity toward right-handedness, an aversion or negative feeling toward left-handedness, and that this right/preferred and left/unpreferred dichotomy manifests itself in their behavior and language. The example of the right/in and right/out door arrangement suggests that this dichotomy affects the business community's decisions about building design. Since time and resources prohibit examination of all of the Ann Arbor business establishments, and because I work for a local financial institution and deal with others, I chose to focus my field of reference on the local FIs to see how this dichotomy affected them. Information concerning relevant aspects of FIs in general is necessary before proceeding further.

According to the United States Savings and Loan League (USSLL), "An association obtains its funds principally from savings accounts and the funds thus obtained are available for investment as the association's charter permits" (USSLL 1950:4). While the USSLL addresses itself to savings and loan associations, this statement applies equally to all types of FIs since, although each FI may focus on different types of investment, they all have the same source of funds—savings.

Therefore, every FI needs to persuade its future customers to relinquish one of their most valued possessions (money) for a period of time, as well as to persuade existing customers to keep their money with the FI. Since the customers cannot actually see their money once it has been relinquished, they must have faith that the FI will see to its safe return at some time. Success in acquiring customers' money, therefore, is directly related to the degree to which customers perceive the FI's trustworthiness; in other words, it depends on the kind of image the FIs project.

A paper by Larry Gard entitled "Banks—A Ritual Institution" quotes bank officials as stating that their institutions are "mainly in business to make profits" (1978), while an FI manual directed to employees states that "With your [employees'] help the bank can grow even stronger and larger." In order to produce a profit or to grow, it is important that customers respond favorably to the FIs' need for their savings. After all, there cannot be investments unless there is something to invest. To help encourage the proper response, FIs rely not only on reputations of efficiency, trustworthiness, and convenient services but also actively strive to affect the customer's perception of the FI. They seek to project an image that they perceive as beneficial to themselves. Various employee manuals illustrate FIs' awareness of the delicate relationship between their survival and prosperity and their image. One manual states that "everything about our appearance makes an impression upon others," another that "public respect is the backbone of the image we have maintained . . . and we feel it vital to our future." A third says

> Our whole organization depends for its very existence on customers from all walks of life. Our . . . customer provides the income which supports the organization and allows it to expand and prosper.

From these excerpts we see that FIs perceive a direct relationship between image and prosperity. Some of the ways in which the FIs' image can be projected are through propaganda, actions, and the buildings that house the institutions. The following example of how First Federal of Detroit (1979:4) describes its new headquarters illustrates how FIs perceive buildings as image projectors.

> The dark-toned granite which sheaths the exterior has a brilliance created by polishing. Contrasting metals are used to give the building dignity and reserve, which [are] called for in this type of structure, the home of a large financial institution.

If the exterior of a building is used to transmit tacit messages such as "dignity and reserve," then it follows that the inside will also be used in the same manner. The messages transmitted through the interior decor, however, have an additional aim. Once the customer enters the building there are usually one or two basic financial decisions which s/he can make. The customer can either put money into the institution's coffers via savings or take money out of the institution by withdrawing funds already in an existing savings account. It follows, based on the profit motive and the desire to expand and prosper which was seen earlier, that the FIs would like to make life more comfortable for the customer who saves while giving "withdrawal" symptoms to those wanting to deplete their savings. Thus the interiors of their buildings attempt to project the savings department as more prestigious than the withdrawal department.

The department that is involved with cash inflow is often referred to as "customer service" (CS). It is up to this department to bring new money into the FI, and it is judged by what kind of savings volume it can generate. It handles most of the new money and large deposits that are funneled into the institution. One of its primary functions is to set up long-term savings accounts where the customer promises to leave her/his money with the FI for an extended period of time, sometimes as long as eight years. This department contains settings where the customer can sit and relax while waiting to be served by a CS employee. When her/his turn arrives for service someone generally indicates which employee s/he is to speak with and quite often the customer is actually escorted to the employee's desk. There the customer can sit and discuss in leisure the particular financial transaction or problem in which s/he is interested.

The department where withdrawals are made is generally called the Teller Area (TA). This department handles the payments made on loans extended to customers by the institutions, cashes customers' checks, takes in deposits and disburses withdrawals from existing savings accounts, and may open special accounts such as "Christmas Clubs." It should be noted that although this department takes in deposits and opens special accounts, these transactions represent a very small percentage of the monies flowing into the FI. These deposits are generally small amounts and are not for extended terms. The money can be deposited today and withdrawn tomorrow.

The physical attributes of the TA are very different from those of the CS. The employees (tellers) who serve the customers do not have desks, but stand at a shared counter. The customer cannot sit and wait for service but must stand in line with fellow customers and wait her/his turn. When a teller is available to give service s/he usually calls out "next." The first customer in line then walks up to the teller, transacts her/his business, and departs. There is not a great deal of privacy for either the customer or the teller, and if the customer has a complicated question or problem concerning finances s/he is generally sent to the CS department where s/he is greeted by name. Thus it is clear that the TA is basically a clearinghouse for day-to-day minuscule transactions.

From these observations of the functions and physical attributes of the CS and TA departments it is apparent that FIs perceive the CS to be the desirable department for customers to frequent and that they have attempted to enhance its image accordingly.

Since the FIs are themselves involved with the same right/preferred-left/unpreferred dichotomy of the society in which they exist, it follows that they would have an awareness, either conscious or unconscious, of its implications and ramifications. Furthermore, since FIs are familiar with projecting tacit messages through the exterior and interior of their buildings, it also follows that this dichotomy would come into play in the projection of these messages. Is there a connection between the FIs' desire to enhance the CS department's image and their cultural perception of the preferred status of right-handedness? An investigation of the sample group of FIs appears to indicate that there is a relationship between the preferred CS/right and the unpreferred TA/left.

Whenever possible the FIs place the CS area to the customer's right-hand side as s/he enters the building through the main entrance. For some institutions, such as Great Lakes Federal Savings and Loan, some confusion might arise as to which entrance should be considered "main." In cases such as this the location of the "night deposit box" is a good determinant for where the institution perceives most of its customer traffic to be. The purpose of the "night deposit box" is to allow customers who cannot frequent the FIs during normal banking hours to make payments on loans and deposits into existing savings accounts. These boxes are placed on the FI buildings where they are most convenient to the maximum number of customers, i.e., where the most traffic normally enters and exits. Thus for Great Lakes Federal Savings, the main entrance is off of their relatively spacious parking lot.

The placement of the CS to the customer's right-hand side and the TA to the left-hand side was found in six of the nine sampled FIs. For the three exceptions there is a physical impossibility of compliance with the CS/right and TA/left layout. It should be noted, however, that even with the exceptions, the right-hand side was shown to be preferred, for in all of these cases the FIs placed both departments on the right-hand side of the room they occupied.

These exceptions, Michigan Savings, Ann Arbor Trust Co., and National Bank and Trust, all share a common problem. They occupy the long, narrow buildings that are the typical older structures in cities and towns all across the country. Since the entrances for these FIs are on one of the narrow walls, the CS/right-TA/left arrangement is not feasible.

Michigan Savings is a recent addition to the Ann Arbor financial scene. It chose to occupy a site in the same block of Main Street as Ann Arbor Bank, Ann Arbor Trust Co., and National Bank and Trust. Perhaps Michigan Savings felt that it would benefit from the customer traffic generated by the established institutions. Whatever the reason, the decision to have offices in this block limited their office space to a very long and narrow layout. However, both CS and TA departments were placed along the right-hand wall, with the preferred CS department up front by the main entrance, so that customers must pass the CS in order to get to the TA department.

Ann Arbor Trust Co. and National Bank and Trust still occupy their original buildings. Over the years, however, each has acquired its adjacent building/s in order to facilitate expansion. Since the buildings along this area of downtown Main Street share common walls with adjacent buildings, it is a relatively simple matter to convert two or more buildings into one business environment by knocking out parts of the common walls. Thus the current total environments of these two institutions occupy almost square proportions.

Both of these FIs, as is the case with Michigan Savings, had an original floor plan that discouraged the CS/right-TA/left layout, and, again like Michigan Savings, both departments were placed along the right-hand wall with the CS first as the customer enters the building. This arrangement persists today, even though the two departments might now be placed across from each other. Several FI officials not connected with either Ann Arbor Trust or National Bank and Trust told me, however, that it would be difficult and expensive to move an existing TA department and relatively simple to move a CS department. If, therefore, these institutions were to decide to place the CS and TA across from each other, it would seem that the logical department to be moved would be the CS. This would leave the FIs with their preferred CS department to their customer's unpreferred left-hand side and the unpreferred TA department to the customer's preferred right-hand side.

This is probably why these FIs have maintained the placement of the CS and TA along the right-hand wall with the CS up front, close to the main doors, even though this arrangement causes some inconvenience to both customers and employees. This inconvenience is the result of distraction caused by the flow of customers entering and leaving the establishment. For National Bank and Trust the problem is compounded by their new, and quite noisy, automatic doors which (by the way) direct the customer to "enter" and "exit" to her/his right.

The right/left dichotomy's effect on the placement of these two departments appears to be an unconscious response by the FIs to their culture. When questioned as to what influenced the placement of these departments, officers from various FIs cited such factors as lighting, work flow, and traffic flow. None of these officers seemed to be aware that the right-hand preference was exhibited by their institution or any other financial institution in the downtown area.

Such devotion to right-handedness as shown in the Ann Arbor area is not a cultural universal. Aram A. Yengoyan (1977) gives an example of how left-handedness was expressed in a border area of Scotland and England. In this area some of the castles that belonged to families noted for their higher-than-normal degree of left-handedness contained spiral staircases with a counterclockwise spiral. However, adjacent castles that did not belong to the left-handed families had staircases built with a clockwise spiral. Presumably, the counterclockwise stairs enabled the left-handed occupants to defend their castles more easily when fights took place on the staircase. Although further study might find such left-handed pockets of resistance in the Ann Arbor area, the preponderance of evidence indicates that this culture places a high value on right-handedness, to the possible exclusion of left-handedness.

It is not clear why Ann Arbor residents exhibit such a high preference for right-handedness. It is clear, however, from the evidence presented in this paper that the local financial institutions understand, albeit unconsciously, the residents' preference for right-handedness, and that this understanding has influenced and assisted them in communicating tacitly with their customers.

Although this project concerns itself with the residents of Ann Arbor and those FIs within its downtown area, I would like to suggest that its scope can be justifiably expanded to encompass a much larger segment of America. This suggestion is based on the atypical transitory nature of the area residents, which has exposed and infused the community with students and other natives of virtually all regions of the nation over an extended period of time.

Terrence Patrick O'Brien did this study of American death customs by observing five funerals in a Detroit area funeral home. His project also included interviews (on legal aspects of corpse disposal and treatment of corpses) with the funeral home director and with college students on their funerary attendance. O'Brien contrasts Americans' unfamiliarity and discomfort with funerals with different customs of other societies. His comparison of cost of funeral, number of mourners, and social position of the deceased shows that, regardless of social class, Americans are uncomfortable with funerary customs. Death, he asserts, is a leveling mechanism in American culture. Finally, O'Brien's data suggest a roughly curvilinear relationship between funeral costs and wealth. Further research may demonstrate that lower-middle-class people spend proportionately more on funerals than members of other social strata. This pattern may be linked to social insecurity of the lower middle class, which is also suggested by Faber's paper (this volume) on tipping behavior.

17 / **Death, the Final Passage: A Case Study in American Mortuary Custom**

Terrence Patrick O'Brien

Death, the final passage, is universally experienced by the living. Biological evolution necessitates individual death for the perpetuation of the species. The universal impact of death evokes diversified reaction from surviving members of the human species. The observances of funeral rituals provide anthropologists with frequent opportunities for probing into the nature of the human experience. In all societies, the issue of death sheds light on many central values governing human lives. Thus, for example, the events surrounding the death of an individual in America are likely to reflect life in American culture.

The native funeral customs of contemporary America are as peculiar as those observed anywhere in the world. In America, death rituals have developed into lavish, expensive displays wherein the corpse is presented publicly, perfectly groomed, for the final time. A variety of costly accessories and artifacts have become integrated into the American funeral ritual, each attempting to create a beautiful memorial. Anthropologists seek cultural values implied in such ritual practices.

Generally, the American funeral seems to be a somber occasion calling for sympathy, restraint, and sorrow. Such highly artificial and restrained behavior is not typical of the funeral rituals in other parts of the world, such as Madagascar and Borneo (cf. Huntington and Metcalf 1979), where funeral behavior can be considered quite rowdy in comparison to the American funeral. Individuals in many parts of the world attend far more funerals than Americans do and thus are likely to feel more comfortable in their attitudes and observances toward the dead (cf. Kottak 1978*a*:381–383; 1980).

The inexperience of the average American in attending funerals may be related to the aura of mystery surrounding death that seemingly exists in American society. This inexperience appears to be evenly distributed throughout all socioeconomic levels. People of all classes feel equally uncomfortable in contact with death. The implication of such an even distribution within the class hierarchy is that death acts as a cultural leveling mechanism in American stratified society.

This study attempts to show, through observations of behavior in funeral rituals from differing American socioeconomic levels, that death indeed is a cultural leveling mechanism. It is essentially a case study of selected funeral rituals aimed at close examination of the behavior of the participating relatives and mourners. Particular attention has been paid to the financial cost of each funeral in an attempt to deduce any relation between socioeconomic class and diversified ritualistic behavior.

A difficulty in studying any aspect of American culture lies in the vastness of the United States. Naturally, such an expansive territory offers a great variety of mortuary customs. Unfortunately, a comprehensive study of funeral rituals occurring in a variety of American geographic locales is beyond the scope of this paper. Thus, this case study is limited to close examination of the events and behavior observed in five Roman Catholic funerals in a suburb of Detroit, Michigan. While five events are hardly sufficient in any statistical analysis, the funerals do exemplify certain generalized features of "the American funeral ritual."

The five funerals were selected to provide the range of ceremonial prices necessary in establishing cross-sectional generalizations with respect to class-specific behavior. Obviously, it is naive to believe that all mourners observing ritual for a specific deceased individual are necessarily of the same socioeconomic class as that individual. Furthermore, although behavior at funerals is undeniably varied, this study also attempts to discover uniform behavior patterns, however general they may be, in funeral rituals of differing costs.

The data come chiefly from observations in each of the five specific funeral cases, along with confidential revelation of the price of each funeral, supplied by the funeral director.[1] Background information surrounding the death of the individual, along with circumstances concerning how funeral arrangements were made, were also provided. These additional pieces of information were crucial in eliminating variables which may have affected behavior in the funeral home.

A survey was also conducted with a dozen college-age subjects in an attempt to establish Americans' inexperience with death and funerals. Also included in the data is an interview with the funeral director or "grievance specialist" presiding over the cases studied.

The behavioral observations were collected by going to the funeral home or, more cozily, "funeral parlor," and inconspicuously examining the actions of those coming to "pay their last respects" in an evening of "viewing" the corpse (an interval of about three hours). Because of the nature of the observations, much of the data collected are more or less impressionistic. Nevertheless, conclusions based on these data are seen as valid on the assumption that the impressions are unbiased.

First, it is necessary to look more closely at the basic morphology of the American funeral ritual and how it can vary in cost. Three price divisions appear to exist on the basis of services rendered ranging from least expensive to most expensive:

1. immediate disposal (pickup of remains directly from hospital or nursing home)
   a. direct cremation:
      $495.00 minimal charge
      $110.00 cremation chest
      $605.00 total expense
   b. burial service:
      $ 495.00 minimal charge
      $ 200.00 "pine box"
      $ 480.00 burial space
      $1,175.00 total expense
2. standard or traditional burial
   price range: $1,200–$9,000
3. elaborate burial (e.g., use of sarcophagus-type coffin and copper-lined burial vault)

The five cases studied involve four funerals which may be classified as standard or traditional and one immediate-disposal-type burial.

1. This and additional information was provided during the author's interview with John J. O'Brien owner/proprietor of Ted C. Sullivan Funeral Home, Detroit.

Focusing on the traditional burial, an itemized account of the services rendered is given in figure 1. It can be seen that the range in prices comes from the selection of casket and of additional related items.

The funeral home that I studied offers a selection of caskets ranging in price from $400 to $7,800, depending on material and craftsmanship. Included in this selection is a three-inch-plank, solid African mahogany model with a copper liner, complete with a glass sheet to cover the top. The copper is covered with an innersprung satin-covered mattress, which offers "perfect posture." This casket sells for $7,815. Not every casket has a silver lining, however; at the bottom of the line is a grey felt-covered wood box with a cloth-covered foam-padded lining selling for $410. Additional related items include the burial or shipping vault, newspaper notice, certified copies of the death certificate, and cemetery charges. The burial vault varies from the basic cement vault ($200) to the "Good Housekeeping approved" model ($1,049), to a Triune copper-lined burial vault ($2,098).

Newspaper death notices are an optional accessory. Checking with the two major newspapers in the Detroit area (11/2/79) reveals that this item costs about twenty-five dollars.

Detroit News	daily: $2.48/line
	Sunday: $2.76/line
Detroit Free Press	daily (1 time): $2.88/line
	daily (2 times): $2.41/line
	Sunday (1 time): $2.77/line

Further checking reveals that a person could run an ad attempting to sell a car for ten days for less than twenty-five dollars. Apparently, the newspaper industry realizes that the grief-stricken family is a captive market. Obituaries, being news items, have no charge, but are run at the newspaper's discretion.

Looking at the minimal charge for a traditional burial (fig. 1), one can see that $250 of the $1,020 goes for the embalming operation, use of the operating room, equipment, and staff embalmer. Embalming is the first part of a sequence of events termed "restorative arts." First, the body is embalmed by replacing blood with three to six gallons of perfumed formaldehyde, mixed with glycerin, phenol, alcohol, and water. The soft interior organs of the chest and abdomen are removed through a long hollow needle with vacuum aspiration and replaced with "body fluid." This procedure, in the language of the embalmers, may be done

Ambulance and two men for removal from residence, hospital, airport	$ 85.00
Embalming operation, use of operating room, equipment, and staff embalmer	250.00
Twenty-four-hour personal services of staff	200.00
Use of funeral home, furnishings, conveniences and equipment, etc.	200.00
Securing necessary certificates and permits	75.00
Funeral coach (local), man, and time	125.00
Family car, man, and time	85.00
TOTAL MINIMUM OPERATIONAL CHARGE	$1,020.00
Casket selected	
TOTAL COMPLETE SERVICE SELECTED	
ADDITIONAL RELATED ITEMS SELECTED	
Vault, burial, or shipping	
Newspaper notice	
Certified copies of death certificate	
Cemetery charges	
Miscellaneous	
Sales Tax	

Fig. 1. Itemized Account of Standard or Traditional Funeral Service

"for show" or "for blow," depending on whether the body will be laid out (for show) or donated to a medical school in the interest of science (for blow). If it is "for show," next comes the restorative phase, wherein mutilated corpses are crudely stitched back together, the scars disguised and missing limbs replaced with plaster imitations. Emaciated corpses are filled out with injected embalming cream, and various features such as hair, teeth, lips, and fingernails are attended to. Finally the corpse is shaved and dressed.

Here you may ask, why are Americans willing to pay $250 for this procedure? Is it required by state law or the board of health? The answer to the second question is an unequivocal "no." Michigan state law requires embalming only if the body will not reach its final destination within forty-eight hours, or if cause of death is a contagious disease (State of Michigan 1978). The answer to the first question appears to be that Americans feel the importance of perpetuating the traditional "three-day burial," in which the corpse is laid out for two evenings in the funeral home and buried on the third morning. Since this lasts longer than forty-eight hours, the body must be embalmed. The body is embalmed in 95 percent of all of the standard burials. It should be pointed out that it is often difficult to lay the body out for only one night, with burial the next day (thus within forty-eight hours), because bureaucratic processing requires a physician's signature certifying death. The death certificate requires the physician to certify that death occurred at the time, date, and place, and due to the cause, stated. If the death was accidental, suicidal, homicidal, or "suspicious" the signature of the medical examiner is required. This processing often necessitates embalming under the provisions of state law.

Turning now to the five case funerals, a summary of relevant physical data is given in table 1. The names of the deceased have been withheld out of respect to the surviving family members and friends, and certain details have been modified to preserve anonymity.

Typically, in Catholic funeral homes the casket has a kneeler, so that mourners may kneel in front of the corpse and "pay their last respects." The family members or surviving spouse typically stand several feet from the coffin and receive visitors. I paid attention to whether mourners proceeded directly to the coffin or first stopped to offer condolences to the family. Several interesting patterns of behavior were observed; in each of the five cases there was avoidance, either of the deceased or of the family members. Some mourners devised elaborate schemes to circumvent the widow or widower and get to the deceased. This avoidance of the grief-stricken family probably reflects insecurity due to lack of experience in offering condolences. After all, what do you say to someone who has just lost someone dear? Certainly nothing that is said will restore life to the deceased.

Other mourners went directly to the family and avoided the deceased. Eye movements revealed that many mourners were unwilling even to glance in the direction of the corpse. Many moved cleverly so that their backs were facing the body. This type of avoidance may be based on a superstition of some kind, along with fear of the dead. It is ironic that many mourners exhibited this hesitance to look directly at the corpse, since, as stated previously, 95 percent of all bodies go through an expensive restoration process.

In addition to these avoidance patterns, mourners' behavior at all of the funerals seemed artificial, uncertain, and clumsy. This may be directly attributable to Americans' inexperience in attending funerals.

TABLE 1 Summary of Data Pertaining to Funerals Observed

	Race	Sex	Age	Occupation	Cause of Death	Cost	Arranged by	No. Cars in Procession	Additional Circumstances
Case 1	White	Male	60	Attorney	Cardiac arrest; coronary thrombosis	$4,025.00	Spouse	28	Overweight; physically unfit
Case 2	White	Female	81	Grandmother	Septicemia	$3,650.00	Son	18	Failing health; placed in nursing home four years prior to death
Case 3	White	Male	32	High school science teacher	Cerebral hemorrhage	$6,100.00	Parents and spouse	82	Death unexpected; passed out and was in coma for a week; autopsy revealed cerebral aneurysm
Case 4	White	Male	70	Retired auto worker	Lung cancer	$2,800.00	Spouse	10	Heavy smoker; emphysema victim; was undergoing chemotherapeutic treatment
Case 5	White	Female	81	Unknown	Coronary thrombosis; apoplexy	$1,240.00 (one night; no limo)	Son with social services	—	Immediate burial

The results of my survey (table 2) revealed that approximately 67 percent of those polled (ages eighteen to twenty-eight) remembered attending fewer than five funerals in their lifetimes. Granted, older mourners have probably participated in many more. But the survey does confirm that Americans are not enculturated into being comfortable with the universal experience of death *when they are young*, and this makes people uncomfortable in dealing with death when they get older and have occasion to attend more funerals. The fact that none of the dozen young Americans polled had attended more than fifteen funerals is probably indicative of larger trends in American society.

There is also an apparent void in the literature on death. Many authors neglect death, while discussing other rites of passage. For example, Victor Turner's *Ritual Process* (1974) deals with just about every ritual transition except death and funerals.

My hours of observation thus revealed remarkable similarities in behavior patterns of the mourners attending the five different funerals. No readily observable difference in behavior existed between the most expensive funeral and the least expensive. All of the mourners appeared to be equally uncomfortable regardless of socioeconomic class (as assessed on the basis of their mode of dress). Although people expressed discomfort differently, none were at ease. These observations appear to confirm the hypothesis that death acts as a cultural leveling mechanism in American stratified society.

An interesting side observation is the relation of the cost of the funeral service to the number of cars in the funeral procession. A linear relationship appears to exist, with the number of people attending the funeral increasing steadily with the cost of the ceremony. This finding would seem to indicate that with increasing status in the socioeconomic class hierarchy, greater importance is placed on attending the burial rite. This pattern is analogous to the findings of Sudnow's (1967) study of dying in a large public hospital in California, which showed that poor people are significantly more likely to be declared "dead on arrival" than are more well-to-do people. "Dead" in this context may be translated as "not worth spending scarce medical resources upon." It should also be noted that in case 5 of this study (the least expensive ceremony), no procession was held and a mere eleven people showed up to say goodbye to the woman (during my three hours of observation).

A final, related observation, supported by the funeral director, is the apparent curvilinear relationship between the cost of the funeral and annual income of the deceased (or arranger). The shape of the curve

TABLE 2 Lifetime Funeral Attendance by Twelve Ann Arbor College Students

No. of Funerals	No. of Individuals Attending
Less than 5	8
5–10	3
10–15	1
More than 15	0

Note: The question, "How many funerals can you remember attending in your lifetime?" was asked in Ann Arbor, Michigan, during November, 1979, of twelve subjects, six of whom were male and six female, ranging in age from eighteen to twenty-eight years old.

would probably be hyperbolic, with the expense of the funeral rising steadily with income until a plateau or "saturation point" is reached. The largest point of inflection in the curve occurs in the middle income range. This is consistent with the fact that the middle class is more likely to feel the social pressure to impress their mourning friends. There is the opportunity to "keep up with the Joneses" to the very end; our high standard of living entails an equally high standard of dying.

In summary, the uniformity of ritualized funeral behavior observed in this cross-sectional study does appear to support the role of death as a cultural leveling mechanism in American stratified society. The relative lack of experience on the part of Americans attending funerals can be a factor contributing to an aura of mystery surrounding death in America. It is this aura of mystery that perpetuates the lavish rite that has become the American funeral ritual.

Mary Jo Larson's paper is based mainly on her systematic observation of verbal and nonverbal behavior in twenty University of Michigan classes during fall 1978. She fills out this study of sexual discrimination in language with information gathered from informal interviews, participant observation, and the relevant sociolinguistic literature. Larson's paper provides admirable evidence for tacit enculturation of traditional American sex roles and male dominance. Her etic research strategy uncovers sociolinguistic patterns and discriminatory behavior that many natives (e.g., many professors) have not recognized.

18 / Social Stratification by Sex in University Classroom Interaction

Mary Jo Larson

In every human culture language serves as a social bond through which most interactions must take place. Language also serves as a way to identify and maintain social strata. For example, in the United States we may be aware of the speech differences between a Detroit black, a midwestern farmer, and a white lawyer. Yet the speech differences between women and men, other than pitch, are not as obvious to the average listener. These differences do exist, however, as is shown by the growing research in this area (e.g., Thorne and Henley 1975). Indeed,

> sexual differentiation of speech is expected to occur whenever a social division exists between men and women . . . that is, universally. [Gal 1978:1]

Even though women are attached to the same economic ranks as men, language differentiation serves to exhibit and maintain male dominance within these ranks. This has been demonstrated on many levels. In children's schoolbooks, males are present seven times more often than females. And when females are mentioned, they often appear as auxiliaries to the males, such as wife, John's mother, Bob's aunt (Graham 1975:57–63). Phrases that refer to "man" ("he speaks up for the little *man*"; "one *man*, one vote"; "Neanderthal *man*") clearly express the concept that males comprise the human race. This usage "contributes to her [woman's] invisibility" (Besmajian 1974:101). This is also illustrated by works on linguistic and other behavior in different cultures, where special sections are labeled "women's language," implying that the language of the culture in general is that of the males (Swacker

1975:77). As Robin Lakoff (1973) shows, the folk concepts of women's language, or how Americans think women speak, reinforce notions of female superficiality and inferiority. Some studies have shown that the mere fact that a speaker is female colors the receiver's impressions of what she says (Swacker 1975). "No matter what women do, their behavior may be taken to symbolize inferiority" (Thorne and Henley 1975:28). Even in words used to refer to women, their inferior status is made clear (Schulz 1975). Obviously,

> there can be little doubt that speech patterns and particular syntactic, semantic, phonological, and intonational structures function to communicate the cultural and social meanings that cluster around sex roles. [Zimmerman and West 1975:106]

In addition, nonverbal behavior is closely linked with verbal behavior in communication. Eye contact, smiling, and touching can all be used to establish dominance or submission. There have been many studies on this as well (see Thorne and Henley 1975:290–303).

In this study, I will explore the hypothesis presented by Zimmerman and West,

> that the distribution of power in the occupational structure, the family division of labor, and other institutional contexts where life chances are determined, has its parallel in the dynamics of everyday interaction. . . . There are definite and patterned ways in which the power and dominance enjoyed by men in other contexts are exercised in their conversational interaction with women. [1975:105]

In order to see if male dominance is established and maintained through patterns of conversational interaction, I have studied classroom interaction at the University of Michigan, a prestigious midwestern university. The school has a good reputation and most of the students come from middle- to upper-middle-class homes. Most students are ambitious, expecting to eventually have power positions in our culture. Attending a prestigious school and establishing intellectual and personal prowess is the first step in that direction, or so they think. Competition is intense.

In gathering data, I visited twenty classrooms, all of which, except two, had no more than thirty students. The study was limited to classes within the School of Literature, Science, and the Arts, where students receive a general liberal arts education similar to that given in any liberal arts college, so that the results would not reflect the bias of special university programs. Since the results from any one class could be due to particular individuals in that class, the classes studied had different locations, teachers, subjects, and students. The classes fell into these general areas: economics, English, history, political science, psychology, anthropology, Afro-American studies, biology, chemistry, physics, philosophy, geography, astronomy, and education. These subjects would be included in any liberal arts catalog. The specific classes were picked from a time schedule and course description list, with the need for a broad range of topics kept in mind. The situations were semiformal, with a teacher (professor or teaching assistant) leading the students in discussion. My role was strictly as an observer; I never participated in the discussions.

Before each class officially began, I counted the number of females and males in the room and noted the gender of the teacher. During the class, I would keep track of the number of times interactions were initiated by students and the sex of the initiator. The amount of time taken by each student was recorded in units of seconds and classified by sex. I also counted the number of times a student spoke back to the teacher in a follow-up interaction after the teacher had responded to the student's first comment or question. The time taken for this interaction was included in the general time category.

I attempted to classify each interaction as one of four possibilities: initiating a new idea, clarifying previous statements of self or others, answering questions, or asking questions. The results here were not as accurate as desired, for the conversations usually went too fast to keep track of them. I also made observations about every class that would not fit into numerical computations, such as the size of the room, the organization of seats, the teacher's and students' appearances, and how the discussion was run. After these empirical observations had been made, I watched for nuances in teacher and student behavior, such as eye contact, smiling, and student reactions. Although these things were not quantified, they contributed to my understanding of male-female interaction. I also talked informally with many students in order to check out my interpretations. Their observations helped illustrate and confirm my ideas at vital points.

There are drawbacks to ethnographic linguistic research like this, as noted by Susan Ervin-Tripp (1973:257).

> The drawback of such studies is that normally there is so much variation at once that we can find descriptive information about distributions but little knowledge of which of the covarying features may be effective.

Many variations from class to class affected gender interaction. I could not tape the conversations, which led to less than perfect accuracy. I could not watch for as many factors as I would have liked. This study does not cover any out-of-class behavior, such as the sexism that may occur on a more personal level between a student and teacher, that contributes to female and male behavior in class. But the research did reveal some general trends that display the establishment of male dominance in a university setting. Once this has been established, other studies can follow focusing on specific problems, such as teacher response to male and female students. All of my analytical conclusions could serve as the bases for further studies. Therefore the results obtained by this project should be considered as tentative.

The results of my observations support other studies done previously in this area: males dominate social linguistic interaction. In the twenty classes, 222 students, or 51 percent, were male; 215, or 49 percent, were female. Yet overall, males spoke 3965 seconds to the females' 1714. Men monopolized 70 percent of the conversation, while women only controlled 30 percent of that time. This is a ratio of 2.3 to 1. Correspondingly, women initiated significantly fewer interactions with the teacher than did men: 127 to the males' 280. Of the total number of interactions that were initiated, 3l percent were by females. Yet when follow-up interactions were counted, only 26 percent of them were by females. Although the difference between 31 percent and 26 percent is not extremely large, it is worth commenting on. Men and women were 11.6 percentage points further apart on follow-up than on initiated interactions. Women did not get as much of a chance to respond to the teacher. In six classes women had no follow-up opportunities, while this happened in only one class for males. Most of the females' speaking time and turn taking took place during their first interaction. Men, in contrast, accounted for 74 percent of the follow-up interactions, while making 69 percent of the initiatory moves.

When the initiated interaction and follow-ups are added together, the total number of turns taken by each sex can be counted. Men took 422, or 70 percent, of the total 599 turns, while women took 177, or 30 percent. This is the same as the percentage of total time taken by males and females. This means that males and females talked, on the average, about the same amount of time per turn (approximately 9.7 seconds). The close proximity between the percentage of turns and the percentage of time taken points out a correlation between turns taken or given and talking time. In analyzing interactional behavior, this shows the importance of the allotment of turns. Those who get the turns dominate the conversation.

Another important factor is how much time men and women spoke for their representative numbers. In coming to an average number, I took the total amount of time spoken by the females and males within a particular class. Then I divided each of those total times by the number of the given sex in the room. That gave the time represented by the sex per person. After figuring out each class, I added all the results for each sex together and divided by twenty. Men, for their representative numbers, spoke twenty-four seconds per person. In contrast, women spoke eight seconds. This is a ratio of three to one. This statistic shows just how unfairly men and women are represented in conversation. In many classes the ratio was even more unbalanced. There was a somewhat more even representation in classes that had significantly more women than men.

Although variation existed between classes, these averages show the general trend. The variations between classes help point out reasons or factors that cause this gross imbalance. These variations will be explored within the context of analyzing and interpreting my observations.

The differing socialization processes that females and males experience (Graebner 1975) result in the patterns of behavior and interaction observed in everyday life, including these classrooms. Very few of the people observed were openly biased against women. The biases are so deeply ingrained that few people realize the extent to which they do exist. In fact, some teachers refuse to believe that there are substantial sexist social pressures that cause women to participate less than men in ongoing cultural interaction at a liberal university like the University of Michigan.

These deeply ingrained biases can be seen in the power play of the classroom. One of the most important factors in determining how much women talk within a given situation is the ratio of males to females. In only one class out of the twenty did the females talk more for their representative numbers (1.25 seconds) than did the males (1.17 seconds). However, in this class there were three and one-third times as many women as men. This was the largest ratio of women to men that I studied and the only one where women spoke more per sex-group member. Conversely, in a class where the men outnumbered the women three to one, the men took up 100 percent of the talking time, resulting in a 57.5 second average per male student in the class. As can be seen from this class, the ratio of men to women is not that important for men (unless it is a women's study course, where women talk about such topics as their anger against men). Men talk despite an overabundance of women. Yet women are significantly influenced by large numbers of men. In only four classes did women account for more than 50 percent of the time taken during student talking. In two of these classes the women still did not speak as much of the time as their percentage in the class: (a) 64 percent of the class, 55 percent of the time; (b) 75 percent of the class, 71 percent of the time. In one class the women's percentage of the class equalled their percentage of time. The other class was the one with three and one-third times as many women as men. In the classes where males dominated in numbers they spoke much more than their numerical representation: for example, (a) 54 percent of the class, 84 percent of the time; (b) 60 percent of the class, 77 percent of the time. This is what happened in most classes.

Women are intimidated by large numbers of males. Women have been accorded a lower status in this country since its inception. They have been considered incompetent in affairs of the world, often being included categorically with children and mental deficients. Any female who has grown up in this culture watching television, reading magazines and books, and going to school is bombarded with the idea that she is not as vital or human as a man. Whether she has rejected these notions or not, when an individual is sitting in a room full of socially acknowledged power holders, she is often intimidated. This does not mean that she believes that men are better, or that the men think they are better. They just hold more power. It is as if five students were put in a room with fifteen professors. The students would be intimidated by the profs even if they believed the profs were no better than they. Yet, put one prof in a room of thirty students and s/he will not be intimidated into not talking by social inferiors.

Another problem with large numbers of males to females relates to the way in which women are conditioned to interact. Hirschmann (1974), who has done studies of single-sex pairs, has found that women tend to support each other in conversation, interspersing their comments with "mn hmm's" and elaborating on each other's statements. Men, in contrast, tend to argue with each other, to participate in "verbal dueling." When women are confronted with verbal dueling they often feel personally attacked. In a class with large numbers of men, women are afraid to enter male verbal combat. From what I could observe, women in classes tend to spend more time clarifying what other people are trying to say than men do. They also answer more questions which have clear-cut, rather than analytical, answers than men.

This stylistic problem was clearly illustrated in one class where the males dominated two to one. One female and two males had to read aloud opinion papers they had written in order to start discussion. The female whispered to the teacher who then announced to the class that the student wanted to stress that it is only an opinion and to be "easy" on her. The two males had no similar qualms. As the female finished reading her paper, accompanied with many apologies, an intense discussion followed. She attempted at various points to reinforce her position, but the males ran the discussion, arguing vehemently. By the end of the hour, the female was practically in tears. She said to her friend, "I'll never get in front of them again." Although the men had meant no harm, and the argument had nothing to do with her personally, she felt assaulted. Women learn not to put themselves out for what they consider to be an attack.

I have also observed that the faster the conversation goes, the more difficulty women have in making their points. In one class, although the females dominated in numbers (76 percent), males controlled 73 percent of the conversation. The males started a very intense discussion in which the females had trouble entering. As suggested by Penelope Eckert (personal communication), women have a hard time fitting into the rhythm of male speech.

Although I did not visit enough classes to observe a definite trend in subject matter and female participation correlations, I do believe that they exist. This would become more obvious in upper-level courses where professional patterns are being set. I would guess, for instance, that female psychology (a traditionally female major) majors talk more than female physics (a traditionally male major) majors. This would be complicated by the fact there are more female psychology majors. Although I observed this trend in some classes, I cannot draw final conclusions from my present data.

The teacher's personality may also make an impact on female participation. However, in order to affect results apart from ordinary variation, the teacher must be either extremely open or obviously closed to female students. Most teachers are neither wonderful nor blatantly sexist, leaving other factors to determine female participation. Of the twenty classes, only one had a teacher who made a noticeable impact because of his personality. Even when I talked to him, I could feel his real interest in my project. He was one of the few teachers who even cared to ask what I was doing and the only one who asked to see the finished project. The students from the class that I talked to expressed warm appreciation and liking for this teacher. Some said he was the best teacher they ever had. In the classroom, males accounted for 59 percent of the students and took up 59 percent of the time. Females, obviously, took up 41 percent of each. The other teachers are not bigoted sexists. The teacher in this case was a warm, truly concerned teacher who respected his students. The females in the class reacted strongly to this acceptance, since women are more sensitive to nonverbal cues (Thorne and Henley 1975:12).

Many social pressures act upon the average teacher to perceive and relate to her/his female students in certain ways which reinforce the women's hesitancy to speak in class. In general, many people, students included, feel that many women are at the university to "catch a man." Indeed, in many cases it may be true, or, as Graebner (1975) states, women go to college "just in case." Professors and students have admitted to me in so many words that this is how they view many women students, especially those who wear makeup and nice clothes. Even female students and teachers believed that women are not serious about their work. This possibility is illustrated by Philip Goldberg (1968) in his study on credibility of females. When told that a set of articles was written by a man (John T. McKay), female students judged the articles "impressive." A similar group of students reading the same articles said to be written by a female (Joan T. McKay) rated the articles "mediocre." This parallels a study by James Whittaker and Robert Meade (1967) which showed that the sex of the communicator was important in establishing credibility of information. When women speak, they are automatically thought to be less serious or credible than men.

This underlying attitude affects the way teachers respond to female students. In one class I visited this became apparent in the eye contact made with the students. Usually I could not tell whom the teacher looked at, because males and females were scattered throughout the room. But in this class, all the males were seated along the back and side of the room. It soon became obvious that the teacher did not even look at the female students; he was concentrating on the males. The females were not even a part of his field of vision. He, at any rate, assumed the female students had nothing serious to say or learn. This same teacher later expressed a real desire to get more women into his department, attributing the lack of females to "the socialization process." Although differential eye contact was not as apparent in other classrooms due to the setups, I observed it to some extent. The most obvious connection to this is where the teacher looks when s/he responds to a student. In some classes, when the teacher responded to a female, s/he would return the focus to the entire class, giving the student no opportunity for a follow-up interaction.

The women I have talked to often reported that they felt put off by professors. Whether it actually happens or not, women feel they are not being taken seriously. The exact reason is unclear, but they feel the message. One female professor I talked to told me of the shock she felt herself when she gave her first lecture and students were actually taking notes. They had to take her somewhat seriously. She became uncomfortably aware at that point that she had not even taken herself seriously. If men do not take women seriously and women do not take themselves seriously, it is no surprise that they talk less in class.

The greatest influence I saw on classroom participation was peer pressure. Most of these students have been socialized in the American culture. When someone steps outside the normally accepted social rules, the group reacts. Linguistic interactional rules are learned. This study shows that there are limitations on how much women are allowed to talk. When any individual woman in this study talked too long, the entire class would begin to react. (By talking too long, I include a number of different turns in which the female is discussing her point. This is not nonstop talking.) I observed this reaction to "overtalkative" females many times over the course of the study, especially in male-dominated classes. First the class would get restless; students shuffled their feet and turned in their chairs. If the woman continued to talk, some students would turn to each other and start small conversations of their own. In a class of about twenty, there would be three or four of these conversations. If the woman still did not stop talking, the students would get uncomfortable, even angry. In one class, one male turned to another and said, "Someone ought to put a plastic bag over her head." A male in the classroom had talked for a longer period of time with no such class reaction.

If all else fails, men resort to laughter to make an overtalkative female shut up. In two classes in this study, two women (one in each class) talked through all of the aforementioned disturbance because they believed in what they were saying and wanted to get their points across. Although they were obviously flustered at the other activity in the room, as was evidenced by a drastic increase in the use of words like "you know," "sort of," and "um" and a frenzied look on their faces, they kept speaking. At that point, for no apparent reason, many of the males in the room started to chuckle. The females then resorted to saying "really" in a whiny voice over and over. The women had said nothing funny. The men's reaction was similar to that old line, "Honey, you're so cute when you're angry." It effectively destroys the seriousness and credibility of what the woman is angry about. If the woman still does not stop talking, as happened in one class, the men interrupt. The other females in the room also react adversely to an overtalkative female, but their reactions are a subdued version of the males'. Their side conversations were quicker; they rarely laughed or interrupted. These techniques worked well in punishing or restraining women who tried to dominate a conversation. Besides working to quiet a woman within a given interaction, this process teaches that female and all other females in the room the lesson of male dominance and the consequences of fighting that dominance. Any woman who has gone through it knows the humiliation and frustration she can feel at the hands of her peers. Usually, however, the process is on a more subconscious level. (In no instances did I see a teacher participating in this "shut up" process.)

As was shown by the study mentioned previously (Goldberg 1968) the sexual identity of a person may predetermine how a group will react to his/her statements. A woman's performance may be perceived as inferior regardless of ability. Women in general are very careful about what they say. In the classes in this study, women generally had well-thought-out questions and statements, while men tended to say or ask anything that came into their heads, as long as they could get a turn. Since there is a bias against a woman even before she opens her mouth, she had better be sure that what she says is worthwhile; otherwise her "stupid remarks" would sound really stupid and serve to reinforce the idea that women are not quite as competent as men. This tendency to talk only when prepared cuts down on the number of turns, and therefore the amount of time, that any woman will take. As women are socialized through television, schooling, and their families to view themselves as social inferiors, it is difficult to establish themselves as intellectual equals of men. Not only do males think women are not as vital or intelligent as they (even though they say they do), women feel it themselves. One male I talked to, who firmly believed in equality of women with men, was surprised, upon reflection, to find that he felt more impatient when a woman said something stupid than when a man did. He felt "Why did she even bother to open her mouth?" Most people seemed to react more negatively to a woman saying a stupid thing than a man. Even if women believe in their full personhood and equality, in struggling to demonstrate their worth they become deathly afraid of saying anything stupid. Their "sisters" put on pressure as well. If we say anything stupid, they will think we are stupid and inferior. Therefore, be careful of what you say. Women have to represent and speak for their whole sex every time they say something. We still have not gotten away from the classifications of people (males) and females. Effectively, this keeps women from participating in discussions as much as men.

All of these factors that keep women quiet lead to one final factor that continually contributes to less participation: women have less confidence than men. When women are accustomed to supportive talking and men use verbal assault, when women cannot get the rhythm of male interaction, when teachers do not include women in their visual field, when women have a stigma attached to them throughout their socialization, when classes shut them up, ignore them, or laugh at them, when they are punished more than men for saying something stupid, it is hardly surprising that women feel less confident about themselves. Men's extra confidence can be observed anywhere on the campus. In fact many men tend to overestimate their capabilities, as is shown by the fact that they apply to graduate schools for which they are not qualified more often than women do.

This male overconfidence and female underconfidence were displayed in one of my classes. Although I was a participant in this class, the example is relevant to this essay. Women in the Residential College view themselves as competent and intelligent and easily the equals of men. The class contains all seniors. There are eleven very assertive women and four men. The women participate extensively in classroom discussion. Yet, when we discussed the prospect of doing a radio talk show on what we had studied, every woman said she did not want to go on the air. All four males, however, said they would be willing to do it. When we discussed why those of us who did not want to do it felt hesitant, we found that all of us (i.e., the females) felt we were not competent enough to discuss our field, that we would do more damage than good in talking from ignorance. Yet, with one exception, every male in the class was much less knowledgeable about his field than the women interested in that subject area. The men felt more confident while being less competent.

During this study, many women exhibited a lack of confidence. Although it was difficult to keep track, women seemed to ask more questions while men more often stated what they thought. Even if women had new ideas, they tended to state them in question form. Also, in two cases where women had to give prepared speeches in front of the class, they exhibited much less confidence than the men. They used more words like "sort of," "um," and "you know." (This did not occur in normal discussion unless they were put under pressure.) The women hesitated more in their speech. Most importantly, they smiled a lot. What they were saying was not funny at all. Among primates, smiling is thought to be used as a signal of submission, nonaggression, or cooperation (Silveira 1972). The act of smiling assures that no aggression or harm is intended. They are perhaps apologizing for momentarily overtaking the male dominance structure. This act of smiling may also create feelings of superiority in the men they are addressing.

The university, like all other educational institutions, serves as a socializing agent. We are learning much more about life than just about mathematics or exotic tribes. All the previous ideas and values set up for the students earlier in life are played out. If a woman, after four years, has still retained her purpose and self-confidence, she may go on to struggle at the next power level in an attempt to join in the valued work of this society. Contrary to what many university professors and students think, women are discouraged from participating. In a sense this is the "real world," a mini-reality. Not only are we encouraged to play out our social roles; these roles and ideas become firmly entrenched. In asking what we can do about language and interaction, Thorne and Henley (1975:30) suggest

> that language and society cannot be so easily separated. Speech is a form of action, not simply a reflection of underlying social processes. To call people Mrs. or Miss is to help maintain a definition of women which relegates them primarily to family roles. To use he or she, rather than he, for sex-indefinite antecedents is a tangible gesture of including, rather than excluding women from consciousness. Males who consistently interrupt females in conversation are engaging in acts of social domination. In short, verbal and non-verbal communication patterns are not simply epiphenomena;

they help establish, transmit, and maintain male dominance. Language change is obviously not the whole story, but it is certainly a part of social change.

Self-Evaluation

Before I began this project, I had never done any field research. By the time I finished, I had gained valuable insight into a particular aspect of American culture as well as the way in which cultural institutions work within society. If the project is approached with time and care, the rewards are great. Once you have seen into the internal workings of our society, your understanding of yourself and your role in your culture will never be the same.

This project optimizes the cultural learning experiences by allowing you to concentrate on an area of interest to you. You should decide on your topic very early in the term, the earlier the better. I found that designing the project was not easy. Once you have decided on an area, you still have to concentrate on a small topic within that area. Once you have found the topic, you have to create a method for collecting data. I spent a lot of time in preliminary observation and readings before I knew how small an area to focus on, and what specifically to focus on. Therefore, I did not have as much time as I expected to have for my data research.

You should try collecting data a couple of different ways before you decide on a specific method. Through preliminary observations, you will see what important things to watch for. You will also be able to set your limits more realistically. I think one of the beginner's biggest problems is trying to deal with too much at once. Although it may seem insignificant at first, focus on one tiny part of what you want to know. You don't have the time to thoroughly study all you would like to within a certain area. But by setting your limits early and setting modest goals for yourself, you will experience the reward of having discovered something concrete that you can share with others. Instead of ending with a handful of unsubstantiated general observations, you will have accurately illuminated a small piece of American culture which you can then fit into the whole.

Also, take advantage of the help offered by the project directors (teaching assistants). They are invaluable in the process of focusing and deciding on specific methods. I was lucky enough to be introduced to Dr. Penelope Eckert, a professor who happened to be interested in my topic and gave me guidance. If you can find a prof who is working in your interest area, use him/her; s/he will probably be happy to help a student doing research on a related topic.

Once you have concrete results from your observations, you can relate them to the society as a whole. Although I could, and maybe should, have written more on the relationship of classroom gender interactions to United States culture as a whole, I have experienced the fact of it many times over. I can see now how power and culture are simultaneously used, taught, learned, and perpetuated. If you put the necessary time and care into the project, you will experience and understand your culture differently from then on.

Van de Graaf and Chinni investigated still another linkup between gender and sociolinguistic patterns. On the basis of formal taped interviews with seven University of Michigan sorority members, Van de Graaf and Chinni tested the hypothesis that women with traditional female majors (e.g., dental hygiene) would be more likely to use the term *girl* when referring to other females than would women with less traditional majors (e.g., engineering), who would be more likely to use *woman*. Analysis of the taped responses to the same twenty-nine questions asked of each respondent did not confirm their hypothesis. Yet, in explaining why their hypothesis turned out to be wrong, Van de Graaf and Chinni's analysis exposes changes in gender terminology and some of the determinants of choice of gender terms.

19 / **Gender Term Use among Collegiate Women**

Tina Van de Graaf
Francine Chinni

Language reflects a people's world view. Thus, studying the language terms that different people use to refer to others, in their close social network and outside of it, reveals much about how they see both themselves and others. One especially interesting aspect of language is gender terminology. It is our opinion, as natives of America, that the term *girl* connotes youth and, therefore, dependency, while the term *woman* connotes age and independence. We expect to find that women who are in traditional female roles will use the term *girl* in reference to themselves and other females, whereas women in nontraditional, more independent female roles will refer to themselves and other females as *women*.

Methods and Procedures

Seven female subjects were interviewed, all of whom live in an undergraduate sorority house at the University of Michigan. These seven women were chosen from a group of forty-five live-in members of the sorority. First, we collected information regarding their major field of study. Then, an attempt was made to equally represent traditional (e.g., nursing, teaching, physical therapy, and other female-dominated careers) and nontraditional (e.g., engineering, medicine, law, business, and other male-dominated careers) occupations in the majors. Francine Chinni, who is not a member of the sorority, selected the interviewees on the basis of their majors, by choosing subjects from the alphabetical roster of the sorority. The subjects chosen ranged in age from eighteen to twenty-one. They were all Caucasian and from middle- to upper-middle-class family backgrounds. The majors of the women chosen were: aerospace engineering, chemical engineering, medicine (cellular and molecular biology as undergraduate major), business, physical therapy, and dental hygiene.

These interviews were taped for later analysis, and there was no time limit placed on the interviews. In all but one case, two interviewers were present when each subject was questioned. Twenty-nine questions were asked, in the same order, of each woman. The questions concerned four different subjects: family background, sorority life, academics and career goals, and women's issues. The questions were asked in the order stated above with the intention of moving from personal to general societal aspects of each woman's life.

Results

In examining the data, various trends became apparent. In all cases, *girl* was used in reference to women with whom the subject was personally familiar. These *girls* were either close to the subject in age or younger, most commonly being sorority sisters or peers. On one occasion, the aerospace engineer used the term *woman* to designate someone in one of her classes, but we have no way of knowing the age or degree of familiarity of the student.

In forty-seven out of forty-eight cases (see table 1) where *woman* was used, it was used in reference to a female with whom the subject was not personally familiar. This term occurred most frequently when the subjects were questioned about women's issues and careers. The dental hygienist did not use *woman* at all, and the aerospace engineer used *woman* exclusively. The term *lady* was employed on two occasions, by the physical therapist and the dental hygienist.

To refer to males, the term *man* was rarely used. In eight out of nine of the instances in which *man* was used, it referred to males in general, people with whom the subjects were not personally familiar. The term *guy* was used most often for males (twenty-three out of thirty-two times).

TABLE 1 Gender Term Use among Interviewees, by Academic Major

Context Number[a]	Aerospace Engineer	Chemical Engineer	Pre-Med	Business	Nurse	Dental Hygienist	Physical Therapist
1	3	7	4	4	14	0	15
2	0	1	0	0	0	0	0
3	0	0	0	0	0	1	0
4	0	1	1	0	3	7	9
5	0	0	1	0	2	0	5
6	0	1	0	0	0	0	0
7	2	0	0	0	0	2	0
8	1	1	1	0	3	4	9
9	0	0	0	0	0	1	1
10	0	0	0	0	1	0	1
11	0	0	0	0	0	0	0

a. Explanation of contexts of gender term use:
1 = *Woman* used to refer to female older than and not personally familiar to subject
2 = *Woman* used to refer to female close in age to or younger than subject, and in most cases personally familiar to subject
3 = *Girl* used to refer to female older than and not personally familiar to subject
4 = *Girl* used to refer to female close in age to or younger than subject, and in most cases personally familiar to subject
5 = *Man* used to refer to male older than and not personally familiar to subject
6 = *Man* used to refer to male close in age to or younger than subject, and in most cases personally familiar to subject
7 = *Guy* used to refer to male older than and not personally familiar to subject
8 = *Guy* used to refer to male close in age to or younger than subject, and in most cases personally familiar to subject
9 = *Lady* used (in intermediate sense)
10 = *Boy* used (referring to children)
11 = *Gal* used

Discussion

The data did not confirm our hypothesis. Women in traditionally female occupations did not use *girl* more often when referring to females than women in nontraditional occupations. Nor did the latter use *woman* more often. While the dental hygienist did not use *woman*, and the aerospace engineer did not use *girl* at all, their uses of *girl* and *woman* respectively were consistent with the other subjects' use of these terms. *Girl* was used in reference to someone familiar by the dental hygienist, and *woman* was used by the aerospace engineer in reference to females in general.

There are several possible explanations for our results. The first involves the age and status of the subjects, who are neither women nor girls. They are still in school, living in the sorority situation, which is reminiscent of the home situation. They have not entered the world of work and are still financially dependent on their parents. All of these factors generate and reinforce feelings of youthfulness. The media also reinforce this feeling, by depicting most women as being twenty-five to thirty years old, an "ideal" age in our contemporary culture. Since people view others in their peer group as they view themselves, it is appropriate for our subjects to refer to their peers as *girls*.

Another explanation is that the subjects could be using *girl* as a term of endearment. In almost every case where *girl* was used, the subject was referring to one of her sorority sisters—someone well known to her, and possibly a reasonably close friend. The term *woman* is impersonal, implying distance or respect. In a living situation such as a sorority, in which everyone is considered a "sister" and an equal, it is unlikely that certain people will be referred to as *women*, while others will be referred to as *girls*. This would tend to destroy the unity and community among members of the sorority. As it is, some of the women in the house are officers, which in itself creates differences in status. These differences may be minimized by calling each other by the same term—*girl*.

The term *lady* was used in an intermediate sense. *Woman* was used to refer to females older than the subjects, in a formal, impersonal sense; *girl* was used to refer to personal friends, peers or, younger children. *Lady*, however, referred to older females with whom the subjects were acquainted. It did not connote the same respect as *woman*, but it was not as endearing as *girl* either. For example, the physical therapist used *lady* to refer to a friend of her mother's with whom she had recently talked. Also, from our sample, it appears that *lady* has become a less common female reference term. It was only used twice in seventy-two references to females.

One final consideration in our analysis of female gender terms must be the wording of the questions themselves. In the questions regarding family background and sorority life, and academic and career goals, we used no gender terms. However, in the questions regarding women's issues, we used the term *women*. This may have prompted the subjects' corresponding use of that term. It should be noted, however, that in the one case where the physical therapist was being questioned about women's issues, she did use *girls* once, and then switched back to her use of *women*. Thus, this last consideration may not be totally relevant to the analysis.

We noticed similar trends in male gender term use. *Boy* was only used twice, both times in reference to a small child. The term *guy* was used in a fashion similar to the term *girl*, as an informal male reference term for males with whom the subject was familiar. In all but one case *man* was used in a formal sense, similar to that of women. However, *guy* was used in a formal sense (four times) more often than *girl* (one time in a similar sense). This phenomenon may be age-specific. A study in which subjects with more varied ages were interviewed might shed more light on this question.

In view of the widespread use of the term *guy*, another possible explanation for the frequent use of the term *girl* emerges. It appears that *girl* has become the equivalent of *guy*, replacing the more traditional *gal*. Of the seventy-two references to females, only one was *gal*. Also, as natives observing the culture, we very rarely see or hear the term *gal*. *Girl* is the closest related informal term, and thus has replaced *gal*.

The data collected from our select sample have revealed interesting trends in gender terminology. The major trends we observed were that *girl* and *woman* were used in a similar manner by subjects with both traditional and nontraditional majors. The same trend was observed for the terms *man* and *guy*. Both *girl* and *guy* were used in informal references, with *guy* used in more formal references than *girl*. *Women* and *men* were used as formal reference terms. There is still much room for further research in this area of linguistics. A more disparate sample in terms of socioeconomic status, race, region, gender, and occupation would yield results representative of the total population.

McClafferty analyzes a group to which he once belonged and which he continued to observe—a high school swim team in Michigan. This selection draws on Victor Turner's (1974) discussion of passage rites, liminality, and *communitas* (see essay 3). McClafferty discusses the swim team's rituals of status elevation and status reversal, and points out that the swim team member's micropassages can be seen as contained in layer after layer of superimposed and connected macropassages; he shows how the liminality of the individual team member is contained within the liminality of the team itself. McClafferty also links the skills and behavior of swimming, particularly its stress on individual achievement and competition, with more general values of American culture.

20 / **Rites of Passage on an American High School Swim Team**

Eric McClafferty

Rites of passage accompany changes in place, state, social position, and age. Victor Turner (1974) has devised a method of examining and describing such rites. This paper will examine one type of rite of passage in American culture. Turner's model will be compared to the structure of this rite; the concepts of this model will then be applied in new ways to help explain the differences of this rite of passage from those examined by Turner. This essay will concern itself with a high school swim team, with particular attention to the concepts of liminality, *communitas*, and the individual achievement of status. A major purpose of this paper is to increase anthropological knowledge by contributing a long-term participant-observer study of an American rite of passage.

A review of Turner's (1974) elaboration of van Gennep's (1960) work (*The Rites of Passage*) is necessary to establish the definitions of the phases that will be discussed throughout the paper. In *The Ritual Process*, Turner defines three stages found in rites of passage: separation, *limen* or margin, and aggregation.

> The first phase . . . comprises symbolic behavior signifying the detachment of the individual or group either from an earlier fixed point in the social structure, from a set of cultural conditions (a state), or from both. During the intervening "liminal" period, the characteristics of the ritual subject (the passenger) are ambiguous; he passes through a cultural realm that has few or none of the

> attributes of the past or coming state. In the third phase (reaggregation or reincorporation) the passage is consummated. The ritual subject, individual or corporate, is in a relatively stable state once more and, by virtue of this, has rights and obligations vis-à-vis others of a clearly defined and "structural" type. [1974:94–95]

The attributes of liminal personae (according to Turner) are that they may: be disguised as monsters; wear only a strip of clothing; go naked; have no possessions, no insignia, no outward sign of rank or role, nothing that distinguishes them from their fellow initiands. They are passive and humble; they obey instructors implicitly and accept punishment without complaint (Turner 1974:95).

Turner (1974:96) describes *communitas* as

> an unstructured and relatively undifferentiated "community," or even communion, of equal individuals who submit together to the general authority of the ritual elders.

He also (1974:97) claims that "communitas emerges where structure is not." He adds,

> Communitas is the being no longer side by side (and, one might add, above and below) but with one another a multitude of persons. . . . Yet, communitas is made evident or accessible (so to speak), only through its juxtaposition to, or its hybridization with, aspects of social structure. Communitas involves the whole man in relation to other whole men. [Turner 1974:127]

In chapter 5 Turner describes the two main types of liminality. The first is the liminality that characterizes rituals of status elevation; this type usually involves as its main constituent the humbling of the passenger on his way up. The second type, the liminality of the ritual of status reversal, occurs when groups or individuals who usually occupy low-status positions exercise ritual authority over their superiors.

Before describing and analyzing the rite of passage with this framework as a reference point, the methodology used to collect the data will be explained and critiqued.

For the most part the data were gathered during the four years (1973–76) I was a member of a high school swim team. I joined the team as a freshman and experienced firsthand the stages that I will describe. I also witnessed the passage of approximately fifty other swimmers through a very similar course of events. Since my high school days

I have remained close to the team in many ways. I have been to meets and practices and have remained friends with the people who still swim. My brother is now a sophomore on the team. During the writing of this paper I went home and saw some practices. I talked with (interviewed) several of the swimmers and even saw members of the team giving "swirlies" (participating in initiation rites) to some freshmen. These participant-observer data may have biases that could well be compounded by the selective recall of the observer and my four-year separation from the events. With this in mind I have tried to eliminate from the study any areas where value judgments may occur. Furthermore, I have done my best to be as objective as possible and I now believe that my continuing connections with the team and my recent visits to the practice sessions have shown me that the memories of the rite of passage that I participated in have not much faded. Even so, the data and conclusions must be evaluated with these caveats in mind.

The swim team is all male, all white, and usually includes forty-five to fifty boys from ages fourteen to eighteen. Obviously the results of the analysis will not be generalizable to the entire American population. However, this study is relevant to the study of American culture for at least three reasons. First, a fair proportion of the people who go to high school in the United States participate in some sport. If their experiences are similar to those of the boys on this team, valuable information might be gained on attitude formation in rites of passage and its effect on the individuals thus enculturated. Secondly, if the characteristics of this rite of passage can be found to be common throughout various rites of passage in American culture, a model somewhat different from Turner's might be applied to these rituals and more learned about the culture, the institutions within which the rites occur, and the people who make up both the culture and the institutions. Finally, by examining a departure from social structure and the eventual reaggregation into a structure we can observe clearly what cultural values are important to the process of status attainment and how they are learned and implemented. In these ways we may learn more about American culture through this study.

The goal or point of aggregation in this rite of passage is becoming a member of the swim team hierarchy. Joining the swim team is voluntary; for whatever reasons, the people who join the team have taken the first step toward completing this rite of passage on their own. This voluntary quality will become significant later when achievement is discussed. The first formal step in joining the team is the first team meeting. The newcomers or freshmen (very occasionally a sophomore or junior) have their first formal learning experience about the swim team. They meet the coach, at a distance; they see the "veterans," and learn about practice times, eligibility rules, and other administrative details. Their final instruction is to be "at the pool with a suit on, ready to practice at 2:10 on Tuesday." The freshmen have begun the first stage of the rite, separation. The symbolism necessary for true separation begins in earnest the first day of practice. Instead of going home from school like the rest of the student body, the swimmers pass into the locker room to prepare for practice. The locker room is a rich source of symbols. It is an obvious mediating sanctuary between the outside world and structure, and the world of the pool, team, and lack of structure, for the initiands. No one who is not a member of the team ventures into the locker room; even the coach rarely, if ever, enters the locker room.

In the locker room the freshmen remove their clothing and thus symbolically abandon the attributes and status that they possessed in the structured society outside the locker room. They put on a swim suit that leaves them nearly naked. This stripping is symbolic of the initiates' lack of possessions. The freshmen are losing whatever identity they had and are moving into a state where they are indistinguishable from their partners in *communitas*. All of the swimmers are male; this eliminates any chance for sexual polarity. They are now part of a group that is referred to by the single title "rookies." Their behavior is passive and humble, and they are prepared to obey any instructor without comment.

As they emerge onto the pool deck for the first time they are like the *tabula rasa*; before them is an experience that will teach them a whole new set of skills and attitudes. The skills and attitudes that they learn will shape more than their swimming experience; they will shape the rookies' lives.

Swimming practices are difficult at any stage in a swimmer's career, but the first high school swimming practices are incredibly hard, especially for someone with little swimming experience (and most rookies have very little). This physical ordeal is a further stripping away of the attributes of the previously held status. The scene of fifty near-naked boys struggling in half-submerged anonymity against cold, pain, and the clock is clearly a scene full of liminal symbolism.

This first practice is especially important in the liminal experience of the rookies, but we are also confronted with the liminality of the entire team in this scene, which is repeated over and over again. It is clear that the swimmers who have emerged from the initial rites of passage are still involved in some aspects of a liminal experience. Possibly the liminality of the swim practice is part of the rite of passage from swimmer to nonswimmer. This would mean that when the initiate began swimming he began not just one rite of passage but two. It is my argument that this is exactly what happens. The rites of passage that are embarked upon when joining the swim team are not merely two but can be a whole series of concurrent passages that can range from a brief status reversal to the career-long passage described above. These passages associated with swimming may also make up parts of even longer rites of passage, like the passage through high school. They might also be compared to or classified as puberty rites, as Fiske (1975) does with football, and thus as a part of the passage into adulthood. Some authors (e.g., Young 1962) see these rites as a mechanism that reinforces social solidarity among these young males; this assertion will be discussed later when the aspects of the period of liminality are discussed.

At some point in the rookie's period of *communitas* he will be given a "swirlie." A swirlie, usually led by swimmers with a relatively high place in the team hierarchy, involves grabbing a rookie and forcing his head into a toilet, which is then flushed. This is part of the destruction of the previous status and a reminder to the rookie that he has not yet reached a place in the hierarchy, for only rarely does a swimmer with a place in the hierarchy get a swirlie. Significantly, this is the first treatment the rookie receives as an individual. Before this he is assigned menial tasks along with his equals in *communitas*. Now, even though he has been "returned to nature," he will often come up with a smile on his face, realizing that he is finally the focus of attention as an individual. The attention that is granted here is a sign that the individual is becoming more important than the group. This shift occurs subtly, but is an extremely important one to observe in studying the characteristics of this rite of passage. *Communitas* has not been lost, since the rookies can still be classified as primarily in that stage. The distinction is that although they entered *communitas* together, they will emerge from the condition individually. So, while the group is still important, it begins to alternate in importance with the individual. The rookie has moved into the grey area between liminality and aggregation that is characterized, in this rite of passage, by the alternating focus first on *communitas* and equality, and then on the individual and achievement. The ritual and the liminality of status elevation have begun.

The mechanisms of status elevation and the subsequent movement into the team hierarchy can be divided into two categories: the physical (i.e., swimming prowess) and the social (i.e., personality attributes, communication skills, communication events, and social and technical knowledge). The movement from liminality into structure has the important attribute of being based on the individual and his achievements. This becomes evident as the mechanisms are studied in further detail. Because swimming is a competitive sport and because improvement can be obtained through practice, the rookie who previously had no swimming skill begins to improve. Improvement is usually attainable through hard work; a commonly accepted belief in swimming is that hard work will result in increased speed. Some commonly heard poolside platitudes are "no pain, no gain" and "when the going gets tough, the tough get going." The value of hard work (and even directed masochism) and its effectiveness in helping one reach the hierarchy are explicitly recognized. The physical measures of movement into the hierarchy are many: e.g., dropping one's time, one's first meet, the first ribbon, the first medal, beating a swimmer who has a place in the hierarchy, winning an event, breaking a record, and winning a varsity letter.

The movement of the rookie into the team hierarchy through the "social" mechanisms of status elevation are less obvious and therefore less easily measured. They are, however, very important. It is possible for an individual with the recognized social skills to move into the status hierarchy even if he has only mediocre physical abilities. This type of entry into the hierarchy is more common than the movement of a good swimmer without communication skills into a high-status position. If one or the other skill is present the person will undoubtedly move into the hierarchy, unless the remaining skill is so poor as to cause the weak swimmer to quit the team or the unskilled communicator to be ostracized. The successful combination of these two skills, however, is what allows an individual to rise to the top of the hierarchy.

The social skills that are necessary for this rise are many. A quick wit and the ability to banter and verbally duel with teammates can be very important to status attainment. The rookie who can joke with or insult a member of the hierarchy at the correct time, with the right audience and in an appropriate way, can take a big step toward a status if the insult is accepted. But if the status holder feels that the rookie has overstepped his limits or is getting a "big head," the rookie may soon be reminded of his position in *communitas* with a swirlie.

Getting a nickname is a sign of high status and individual recognition of achievement. Redeye, Captain Hook, Sode, Yogi, Squid, Woody,

J.B., Jasp, Loop, and Domino were names that separated individuals from the *communitas* of the rookie forever. The knowledge of two types of special communicative knowledge is important to status elevation. The first type of knowledge is the swimming argot. There is a whole language of swimming terms and phrases that a rookie could not know. Along with this argot of swimming terms are the historical references and special terms that occurred or were coined in previous years, and about which no rookie can know until instructed.

The jargon related to swimming is a good example of a structural mediator between the physical and social spheres of status elevation. This way of communicating refers almost exclusively to knowledge and skills associated with swimming prowess. When swimmers use words like *skips* (swim, kick, pull, swim), *grab* (a type of start), *split* (the time of a portion of a race), *die* (to barely finish a race), *DQ* (disqualify), *killers* (one lap sprints), *reps* (repetitions), or *fly* (the butterfly stroke), they are referring to swimming skills, but they are doing it in a social way, through language.

The other type of swimming communication can be classified in four ways: as popular expressions, cheers, historical references, and insults. "Drugs" was a popular expression which meant *bad*, as in Q: "How did you feel when you took last in the five-hundred freestyle?" A: "Drugs, man, drugs." The team cheer takes quite a while to learn; it goes

> acka lacka ching / acka lacka chow / acka lacka ching ching chow chow chow / boomalacka boomalacka sis boom bah / our team, our team rah rah.

Historical references are the overtly expressed recognition of the shared experiences of the team members. The fabric of team history is woven with the mixed occurrences of events of the period of liminality and the individual accomplishments of the team members.

This type of special knowledge separates the liminal swimmer from the swimmer who has a place in the hierarchy. But the knowledge in itself does not determine the individual's place in the hierarchy. The swimmer's position in the presentation and discussion of this knowledge is one of the social determinants of the place the individual will assume in the hierarchy. This is where communicative and social skills become important. The individual with a quick wit or highly personable manner will rise in the hierarchy if the use of these skills occurs in appropriate situations. If the person is accomplished at these skills he learns the appropriate situations for the uses of the above mechanisms of status elevation quickly. Examples of the importance of the position taken in the presentation of this information to the position in the hierarchy are many. The team captain, obviously a member of the upper level of the hierarchy, starts the team cheers and helps teach the initiates the cheers. His presentation of this knowledge helps define his position in the hierarchy. Increased skill in this area may be acquired through practice and observation. Thus social skills and the status position they may bring can also be achieved.

Another component of the social hierarchical system and its manifestation and clarification of position are the group designations that occur within the hierarchy. Whom one spends time with reciprocally determines and is determined by one's position in the hierarchy. A good example of this is the group eating patterns of the post–swim meet meal that is eaten at the "home" McDonald's. With some exceptions, caused by time of arrival, the placement of individuals radiates in a circular pattern of decreasing status from a center established by the higher members of the status group.

It is clear that the move from liminality to aggregation is based on individual achievement. This is in conflict with the claim that Turner makes that this movement is seen around the world to be controlled by "more than human powers." The most important aspect of this is the overt recognition of the swimmers and the quick recognition by the rookies that working hard to move up the hierarchy is an accepted and encouraged practice. Competing with their brothers in *communitas*, and thus moving out of *communitas*, is the beginning step of their reaggregation.

A second type of liminality discussed by Turner (1974) occurs on the swim team; this is the liminality of the ritual of status reversal. Turner (1974:167) describes this as

> points in the seasonal cycle [during which] groups or categories of persons who habitually occupy low status positions in the social structure are positively enjoined to exercise ritual authority over their superiors; and they, in their turn, must accept with good will their degradation . . . often accompanied by robust verbal and nonverbal behavior in which inferiors revile and even physically maltreat superiors.

On the birthday of the captain of the swim team, the other team members line up in the water down the length of the pool facing the pool wall, leaving between themselves and the wall one pool lane open for the captain to swim in. Each swim member has a kickboard. With these boards they begin to push the water in the lane back and forth; the combined movement of fifty kickboards creates turbulence though which the captain is required to swim two laps of the butterfly. This ritual of status reversal is seasonal and it is physically abusive. Another characteristic of the rite which contributes to its liminal qualities is the masking function (cf. Turner 1974:172). The swimmers in the water are not regarded as individuals since they are in the water, dressed similarly, disguised by turbulence, and unidentifiable by the captain who is doing what he can to continue breathing. Another aspect of the liminality of the rite is the equality of the group. The only status distinction being made is the group's superiority to an individual who is normally considered their superior.

Turner says that rituals of status reversal reaffirm the order of structure; the stress in this swim team status reversal is on the *temporary* reversal of status. Everyone is made aware that their superior is lower than them all for a short time. This realization must be accompanied by the recognition of the captain's status before and after the temporary reversal. This recognition, along with the participation of the team in a superior status, reinforces the existing structure while seeming to deny it. According to a Lévi-Straussian structural analysis, the contradiction that is inherent in this status reversal is resolved when the group takes on the characteristics of its opposite in an "inversion" of characteristics.

A final type of liminality that can be observed on the swim team is institutionalized liminality. This is the liminality that is incorporated in the swim team and its modes of operation. Instead of a passing liminal experience we can see characteristics of a continuing liminality. An example of this is the equality and the anonymity that are imposed upon the team by other teams and other observers. The daily act of swim practice has many elements of liminality. The whole team is nearly naked, everyone suffers the same pain, and all are forced to obey a superior. The whole team goes through the transition from the status of the outside world to the propertylessness of the pool deck.

It is interesting to note that the position in the swimming hierarchy is not determined by the position of the individual and his family in the social and economic structure of the "real world." The indirect effects of these positions on the swimmer, such as the ability to communicate, are present and do indirectly affect the rise of the swimmer in the hierarchy. It is my assertion that status groups which enculturate their children with the Puritan work ethic and similar types of economically based achievement orientations and imbue their young with the importance of the individual, which Arensberg and Niehoff (1975) and Kottak (personal communication) all seem to agree is a valid American myth, would be found to predispose their children to success in this type of structure.

As the rookies learn the criteria and methods for removing themselves from liminality, they are also taught (or the belief is reinforced) that it is desirable to aggregate themselves with the members of the structure and to attain status and therefore prestige and power over others. With position comes respect from peers for arriving at a status through individual achievement. This atmosphere creates an environment for competition among members of the structure. This competition leads to further liminality as the achievement orientation creates changes in the structure, and so the ritual of status elevation continues. A corollary of this process (which remains unmentioned by Turner) is the ritual of status descent. As one moves down the hierarchy, status is temporarily suspended because of a lack of achievement; this continues until a new status is established. Fiske (1975) describes the cultural function of competition and individual achievement, which prepare

> the novice for activity once again in the society, by concentrating on the skills and performance of the individual player; competition within the group is high—this is a relearning process to prepare for reintegration into competitive society.

One cultural function of *communitas* and institutional liminality is their ability to ease cultural tensions that build within the competitive atmosphere of the team. *Communitas* partially resolves this tension in its equality and anonymity.

In lecture, Kottak commented on the American penchant for joining voluntary associations. We have seen that joining the swim team is a voluntary process. Is it possible that individual Americans seek out voluntary associations to find opportunities for *communitas* as well as opportunities for achievement because the desire to attain *communitas* (which Turner comments on) cannot be fulfilled easily in the larger American culture? The data presented here merely suggest, rather than confirm, this hypothesis.

This essay has examined the importance of the role of the individual and achievement in an American rite of passage. These roles have been seen to have been important in the movement from the second phase of the passage, liminality, to the third stage, aggregation. Another important observation has been the recognition of a series of passages from *communitas* to structure; this temporal series has been seen to have been contained in part by a continuum of liminal experiences which range from the first practice session to the swimming career as a whole. It has been suggested that these micropassages can be viewed as being contained in layer after layer of superimposed and connected macropassages, and that these macropassages continue through time.

Content Analysis of the Mass Media: Sex-Role Stereotyping and Other American Values

Essays 21 through 25 resume the analysis of the mass media, using techniques other than those that predominated in Part 2. The student research projects of Rentz, Hesseltine, and Hill are detailed content analyses of magazines, television commercials, and televised soap operas, respectively. Professor Margolis's paper, which demonstrates how an American myth or ideology can have far-reaching results, is the only selection by a professional. Its inclusion sets the stage for the discussion by Rentz and Hesseltine of media portrayal of sex roles and for Hill's analysis of unrealistic depiction of blacks in television soap operas. Combined with the essays by Larson (18) and Van de Graaf and Chinni (19), these analyses by Margolis, Rentz, and Hesseltine show that many of the same cultural themes, often involving discrimination against women, show up in a variety of contemporary settings. The final chapter is a research paper by Venezuelan student Fermin Diez. As the only nonnative contributor to the volume, he provides a valuable outsider's perspective on an aspect of contemporary American culture—the tremendous popularity of team sports, particularly baseball and football.

Here Professor Margolis shows how a "Blaming the Victim" ideology, first named by William Ryan (1971), is applied to American women, contributing to society's perception and treatment of them as second-class citizens. Such an ideology is universally encountered in stratified societies; those who are deprived of equal access to resources of strategic and social value are themselves blamed for their misfortune. In this way, social attention is diverted from the true causes of inequality and discrimination, and from the need to solve these social problems. Margolis examines manifestations of the Blaming the Victim ideology as applied to women in several areas of contemporary American life. In the world of work, for example, women are blamed for their lack of advancement, when jobs and resources are in fact limited, and when society fails to provide adequate facilities to release parents from many essential responsibilities. Women's health problems are often seen as psychological rather than physical, and women thus receive less careful medical attention than men. Women's depression during menopause is attributed to psychology rather than to objective conditions, e.g., hormonal changes or desertion by a husband who has "run off with a younger woman." If mothers stay at home, they "smother their children"; if they have external jobs, they neglect them. Victims of rape and wife abuse are seen as enticing or aggravating their assailants. Margolis's analysis points up many more applications of this ideology to American women. The dimensions of enculturation that she examines are not just tacit ones, in that many contemporary women have recognized discrimination in particular areas. However, the sum total of the uses of the Blaming the Victim ideology against American women is probably more than most readers suspect.

21 / **Blaming the Victim: Ideology and Sexual Discrimination in the Contemporary United States**

Maxine Margolis

> *Women are a problem not only as individuals, but collectively as a separate group with special functions within the structure of society. As a group and generally, they are a problem to themselves, to their children and families, to each other, to society as a whole.*
>
> Lundberg and Farnham
> 1947, p. 1

> *Only an equal society can save the victim from being the victim.*
>
> Gloria Steinem
> PBS Program on wife beating
> May, 1977

It has long been a cardinal rule of anthropology that one of the main functions of a culture's social and economic structure is the creation of ideologies that perpetuate, or at least do not threaten, the status quo. The need for system-maintaining ideologies is particularly acute in stratified societies where the divisions between haves and have-nots always present potential challenges to the established order.

Blaming the Victim is one such system-maintaining ideology. It helps to preserve the status quo in the United States and other stratified societies by attributing myriad social ills—poverty, delinquency, illegitimacy, low educational attainment—to the norms and values of the victimized group, rather than to the external conditions of inequality and discrimination under which that group lives. According to William Ryan (1971:xii), who was the first to recognize and label this phenomenon, Blaming the Victim is "an ideology, a mythology" consisting of a "set of official certified non-facts and respected untruths."

The primary function of this ideology is to obscure the victimizing effects of social forces. Rather than analyzing the socially induced inequalities that need changing, it focuses instead on the group or individual that is being victimized. This results in distracting attention from the social injustice, thus allowing it to continue. To change things, according to the ideology, we must change the victims, rather than the circumstances under which they live.

In American society this ideology is most often applied to minority groups, particularly blacks. It is used to "explain" their low socioeconomic status, their "aberrant" family structure, and their general failure to reap the benefits of all—it is said—this society so freely offers. Then too, Blaming the Victim is a convenient tool used to account for the underdevelopment of the third world. According to the ideology, underdevelopment is due to some defect in the national character of the nations affected, to their people's lack of achievement motivation or openness to innovation.

Here I will argue that in the contemporary United States Blaming the Victim is also used widely to rationalize the continued economic, political, and social inequality of women. The application of this ideology to women is somewhat problematic in that, unlike other minority groups, women, at least until recently, have not regarded themselves as the objects of collective victimization. Moreover, through the process of socialization, most women have internalized victim blaming—they blame themselves and other female victims for their economic, social, and political problems. Essentially they ask: "What am I doing to make people discriminate against me?"

In writing this paper, I soon realized that women are blamed for a host of society's ills which, strictly speaking, only indirectly victimize women. Rather, in such cases, the victims are their children, their husbands, and their close associates who, it is said, are damaged by female behavior. The best known example of this type of thinking is the claim that women who work neglect their children and, therefore, are entirely responsible for whatever emotional and behavioral problems arise in their offspring. But this is simply a new twist of the victim-blaming mentality since its function is the same: to obscure the current social order's role in creating all manner of social and psychological problems by placing the blame where it is often not warranted.

Then too, it is sometimes difficult to distinguish the tendency to blame the female victim from outright misogyny. Is, for example, Philip Wylie's (1955:chap. 11) charge that men and boys are infantilized by their archtypical "Moms"—whom he describes as women who are "twenty-five pounds overweight," who have "beady brains behind their beady eyes," and who spend their time playing bridge "with the stupid voracity of a hammerhead shark"—simple misogyny or does it have an element of the victim-blaming mentality in it? It is clear why the line between the two is often blurred: it is far easier to victimize a group whom you dislike. By defining women as inferior, less trustworthy, more emotional, and less motivated than men, mistreatment or, at least, unequal treatment is justified and the status quo is preserved.

> *It is not particularly important to a great many working women whether or not they earn as much as men, or have equal opportunities for training and promotion. [Smuts 1974:108]*

Blaming the female victim finds its widest application in the world of work. Here it comes in a variety of guises and is used to "explain" why women are paid less than men and have fewer opportunities for occupational advancement. Women, it is said, work only for "pin money" since they have husbands to support them and, therefore, do not really "need" their jobs. Similarly, it is claimed that women have higher rates of absenteeism and job turnover than men do, along with less interest in moving up the career ladder. These purported characteristics of the female labor force are then used to rationalize the fact that women are overwhelmingly confined to low-paying, tedious, dead-end positions.

According to the "pin money" argument, men must provide for

their families, while women only work to supplement their husbands' income or for pocket money to buy "extras." Using this logic, employers rationalize paying women low wages on the grounds that they do not need their earnings to live on. And, lest it be thought that this justification for salary discrimination has succumbed to more enlightened thinking in this era of the Equal Pay Act and the feminist movement, the comments of a county commissioner in Utah should lay such hopes aside. When asked to explain why male employees had received a 22 percent wage increase and female employees a 5 percent increase, he replied: "We felt that with their husbands working, the ladies could stand the squeeze a little better" (quoted in *Ms. Magazine* 3 December, 1975).

This reasoning is specious since it misinterprets why women enter the labor force: the reasons are overwhelmingly economic. Of the nearly 38 million women employed in 1976, 84 percent were the sole support of themselves and their families, or were married to men whose 1975 incomes were under $15,000. Women's median contribution to family income was 40 percent, with 12 percent contributing one-half or more. Moreover, the only reason many families are able to maintain a middle-class standard of living is that they have two incomes. "Women flocking to work account for the vital margin between solvency and insolvency," says economic analyst Eliot Janeway (1977:66).

One of the most pernicious results of the pin money myth is the failure to take high levels of female unemployment seriously. The belief that women's jobless rates are less worrisome than those of "household heads" again belies the fact that most women work not for pocket money, but because they are the sole support of their families or because their earnings make up a substantial proportion of their household income. By ignoring these facts, a delegate to the 1976 Republican National Convention could pooh-pooh high unemployment rates.

> The unemployment rate tells a dangerously false story for which women are particularly to blame. It's not an economic problem. It's a sociological problem. [Quoted in Porter 1976]

Another component of the Blaming the Victim mentality in the world of work is the purported tendency of women to have higher rates of absenteeism and job turnover than men. These supposed liabilities of employing female labor also have been used to justify lower wages for women as well as employers' reluctance to promote them to more responsible, better-paying positions. Here, it is argued that women are not attached to the labor force, that they just "up and quit their jobs" to get married, or, if already married, to have babies. Why then, it is asked, should employers invest in expensive job-training programs for women or allow them to take on positions of responsibility?

Once again, the facts are ignored by the victimizers. A Department of Labor study of job turnover over one year found that 10 percent of male workers and 7 percent of female workers had changed jobs during that period (U.S. Department of Labor 1975). The number of women who leave work when they marry or have children has declined in the last two decades, and even with breaks in employment, the average woman now spends twenty-five years in the labor force.

It is also claimed that women miss work more than men do since they are subject to "female problems" and are more likely to stay home under the pretext of one minor ailment or another. Here too, the facts speak to the contrary. A recent survey by the Public Health Service found little difference in absentee rates due to illness or injury; women averaged 5.6 days annually and men averaged 5.2 days annually (U.S. Department of Labor 1975). Moreover, women over forty-five had a lower absentee rate than did men in the same age bracket.

Although ideas have changed since the early years of this century, when menstruation, pregnancy, and menopause were viewed as serious illnesses that disabled women and made them ill-suited for paid employment, many hiring and promotion policies still view women as baby makers who, if they are not pregnant, will soon become so. This assumption then becomes the employer's rationale for passing over women for promotion. Nor does the situation improve for older women since the belief that menopausal women suffer emotional disturbances is often used to justify denying them good jobs.

In a similar vein, Dr. Edgar Berman, a member of the Democratic party's Committee on National Priorities, received widespread publicity in 1970 when he questioned women's ability to hold certain responsible positions due to their "raging hormonal influences." "Take a woman surgeon," said the illustrious doctor, "if she had premenstrual tension . . . I wouldn't want her operating on me." Of course, Dr. Berman ignores the research which suggests that men have four- to six-week cycles that vary predictably and also seem to be caused by changing hormonal levels (quoted in Corea 1977:98–99).

Victim blamers also assert that women don't get ahead in their jobs because they lack the ambition to do so. They claim that women don't want promotions, job training, or job changes that add to their work load: "What they seek first in work," says sociologist Robert W. Smuts

(1974:108), "is an agreeable job that makes limited demands. . . . Since they have little desire for a successful career," Smuts continues, "they are likely to drift into traditional women's occupations." George Meany offered a similar rationale in commenting on the lack of women on the thirty-three-member executive council of the AFL-CIO: "We have some very capable women in our unions, but they only go up to a certain level. . . . They don't seem to have any desire to go further" (quoted in *Ms. Magazine*, July, 1977).

Data regarding women's purported lack of ambition are difficult to come by, given that relatively few women have been offered positions of responsibility in the business world. Nevertheless, there is no evidence that the 5.1 million women who held professional and technical jobs and the 1.6 million who worked as managers and administrators in 1974 performed any less ably than men in comparable positions.

Yet another assertion made by Blaming the Victim ideologues is that women are "naturally" good at tedious, repetitive jobs; that they have an aptitude, if not an affinity, for typing, filing, assembling small items, packaging, labeling, and so forth. This view is clearly spelled out in a pamphlet entitled *The Feminine Touch* issued by Employer's Insurance of Wausau.

> The female sex tends to be better suited for the unvarying routine that many . . . jobs require. Women are not bored by repetitive tasks as easily as men.

It was also echoed by the chief detective in the notorious "Son of Sam" case who, in a *New York* article, was quoted as saying that he sent two female detectives to the hack bureau to go through tens of thousands of licenses because they were "judged better able to withstand such drudgery than men" (quoted in Daley 1977).

These stereotypes lack any data to back them up and are simply rationalizations that allow men to assign women to such tasks without guilt. They also help justify the continued ghettoization of women workers in certain "appropriate" female occupations where their purported aptitude for tedium can be put to good use.

> *Many women . . . exaggerate the severity of their complaints to gratify neurotic desires. The woman who is at odds with her biological self develops psychosomatic and gynecologic problems. [Greenhill 1965:154, 158]*

Psychiatry and gynecology have provided lucrative settings for victim-blaming ideologues. Blaming the female victim is the unifying theme in the perception and treatment of such medically diverse spheres as depression, childbirth, contraception, abortion, menstruation, and menopause. The common thread in all is that women's psychological and medical complaints are suspect, that they exaggerate their ills to get attention, and that most female ailments are of a psychogenic rather than a biogenic origin.

Most psychotherapists, wittingly or unwittingly, ignore the objective conditions under which female neurosis and depression arise, and help maintain the sexual status quo by suggesting individual rather than collective solutions to female discontent. The patient is encouraged to think that her depression, her neurosis, is unique, that they are conditions of her own making.

Nowhere is the Blaming the Victim syndrome more evident in the profession than in the diagnosis and treatment of female sexual problems. Lundberg and Farnham, in their misogynist tome *Modern Woman: The Lost Sex*, claimed that the failure of women to achieve sexual satisfaction is a neurosis that stems from a negative view of childbearing and from attempts to "emulate the male in seeking a sense of personal value by objective exploit" (1947:265). Similarly, Freudian psychoanalyst Helene Deutsch (1944) believed that frigidity in women resulted from nonconformity to the feminine role.

Blaming the individual woman for emotional problems that in many cases are related to her fulfillment of traditional, socially accepted female roles obscures the dilemmas inherent in these roles and relieves society of responsibility for her unhappiness. The psychiatrist Robert Seidenberg suggests that the housewife-mother role often gives rise to emotional problems in women who adhere to it. He found that the "trauma of eventlessness"—that is, the absence of stimuli, challenges, choices, and decision making, which characterizes many women's lives—can threaten their mental well-being as much as physical danger (quoted in Sklar 1976).

When women are used as guinea pigs—as in the case of the birth control pill and other contraceptive devices—their complaints of side effects are often dismissed as the reaction of neurotic females. For example, depression, a fairly common side effect of the pill, is discussed in a medical text in these terms:

> Recent evidence suggests that a significant number of these depressive reactions are due to an unrecognized and deeply rooted wish for another child. [Ciriacy and Hughes 1973:300]

The fact that the development of birth control pills and other contraceptive methods has been largely aimed at women is the result of a number of assumptions made by the largely male research establishment. Not only do they believe that conception control is the responsibility of women, they fear the untoward effects of interfering with the male sex drive. Having a healthy supply of sperm is more important to men than ovulation is to women, claim these authorities.

Although the medical profession encourages women to employ problematic contraceptive techniques, it is far more reticent about permitting them to undergo early, medically safe abortions. The reasons are often of the victim-blaming ilk. A staff physician at a county hospital in Milwaukee compared abortions to such cosmetic procedures as face lifts and breast enlargements. "Women know what makes them pregnant and they should have responsibility," he is quoted as saying (quoted in the *Milwaukee Sentinel*, July, 1976).

Even pregnancy and childbirth do not escape the net cast by the medical victim blamers. Morning sickness, for example, is described in one gynecological text as possibly indicating "resentment, ambivalence, and inadequacy in women ill-prepared for motherhood" (quoted in Corea 1977:77). Thus, a condition that is experienced by 75 to 80 percent of all pregnant women, and seems to be related to higher levels of estrogen during pregnancy, is dismissed as a psychosomatic aberration. Others have claimed that many women exaggerate the pain of childbirth: "Exaggeration of the rigors of the process is self-enhancing and . . . affords a new and powerful means of control over the male," say Lundberg and Farnham (1947:294).

Menstrual cramps also are suspect. One gynecologic text writer (Greenhill 1965:154) argues that they often "reflect the unhealthy attitude toward femininity that is so predominant in our society." Other medical texts adopt a similar view. One attributes menstrual pain to a "faulty outlook . . . leading to an exaggeration of minor discomfort," while another states "the pain is always secondary to an emotional problem" (quoted in Lenanne and Lenanne 1973:288).

Victim blaming by the medical establishment reached a crescendo during Senate subcommittee hearings looking into unnecessary surgical procedures. There, the highest ranking staff physician of the American Medical Association argued that hysterectomy is justified—though the uterus was healthy—in women who feared pregnancy or cancer. The chief of obstetrics and gynecology at a Rhode Island hospital agrees: "The uterus is just a muscle" and "It's a liability after children are born because it's a cancer site." The doctor added that his reasoning "ideally" applied to breasts as well, but "this would be a hard concept to sell in this society" (quoted in *Ms. Magazine*, November, 1977). One fact little noted in these discussions is that while it is true that hysterectomy eliminates the possibility of later uterine cancer, the death rate from uterine cancer is lower than the mortality rate from hysterectomies.

Menopause is another medical area in which victim-blaming health practitioners have had a field day. The common medical depiction of menopausal women as aged hags suffering from hot flashes and severe depression has been adopted by the public at large. A judge in Toronto, for example, dismissed the testimony of a forty-eight-year-old woman, stating: "There comes a certain age in a woman's life . . . when the evidence is not too reliable" (quoted in *Ms. Magazine*, July, 1977). This stereotype overlooks the fact that only between 20 and 30 percent of the female population have such symptoms. Moreover, it is usually assumed that depression is caused by the loss of reproductive capacity, while little attention is paid to the objective life conditions of many middle-aged women—their "empty nests," their husbands' inattention, their lack of challenging employment opportunities, and society's glorification of female youth and beauty. Surely these conditions do much to account for depression in middle-aged women. But rather than question traditional sex roles, or the unequal distribution of power between men and women in our society, the medical establishment appeals to the "empty uterus" as the source of female discontent.

> *Whether they like it or not, a woman's a sex object, and they're the ones who turn the men on. [Judge Archie Simonson, Dane County, Wisconsin, 1977]*

Nowhere is victim blaming more pernicious than when it is used to rationalize sexual and physical aggression against women. The courtroom statements of Judge Archie Simonson of Wisconsin show that Blaming the Victim is still too often the norm in the perception of rape and the treatment of its victims. This is also true of wife beating. In fact, attitudes toward abused wives and rape victims are strikingly similar; just as the rape victim is supposed to be an irresistible temptress who deserves what she got, so, it is said, the abused wife provokes her husband into beating her. Then too, it is said that women secretly enjoy being beaten, just as they are supposed to be "turned on" by rape.

There are two components to victim blaming as a rationalization for rape. For one, it is assumed that all women covertly desire rape, and, for another, that no woman can be raped against her will, so that forcible rape doesn't really exist. In combination, these assumptions lead to the conclusion that if a woman is raped, she is at fault, or, as Brownmiller (1976:374) says: "She was asking for it" is the classic rapist's remark as he "shifts the burden of blame from himself to his victim."

Victim precipitation, a concept in criminology often used in rape cases, tries to determine if the victim's behavior contributed in any way to the crime. While an unlawful act has occurred, goes the argument, if the victim had acted differently—had not walked alone at night or allowed a strange male to enter her house—the crime might not have taken place. This point is illustrated by a court case in California in which the judge overturned the conviction of a man who had picked up a female hitchhiker and raped her. The ruling read, in part:

> The lone female hitchhiker . . . advises all who pass by that she is willing to enter the vehicle of anyone who stops, and . . . so advertises that she has less concern for the consequences than the average female. Under such circumstances, it would not be unreasonable for a man in the position of the defendant . . . to believe that the female would consent to sexual relations. [*New York Times*, July 10, 1977]

Another example of this mentality is the minister who wrote in a letter to "Dear Abby" that a young girl whose father had sexually abused her had "tempted" him by "wearing tight fitting, revealing clothes." In light of these opinions, which reflect the deeply ingrained notion that women provoke rape by their behavior and dress, it is little wonder that rape victims often agonize over what they did to cause themselves to be raped.

These attitudes are also evident in the way rape victims are handled by the courts and the police: the victim is more often treated like the criminal than is the rapist. Some states still permit testimony about the victim's prior sexual experience and general moral demeanor, and Brownmiller (1976:419) cites a study of the jury system which reported that in cases of rape "the jury closely scrutinizes the female complainant" and "weighs the conduct of the victims in judging the guilt of the defendant."

Similar attitudes are reflected in a California police manual which states that "forcible rape is the most falsely reported crime," and Brownmiller (1976:408) notes that many police assume that rape complaints are made by "prostitutes who didn't get paid." If a woman is raped by a stranger, the charge usually is taken more seriously than if she is raped by a man she knows. The latter, the police claim, is a "woman who changed her mind."

No matter how women behave in rape cases they are still held responsible for the outcome. While popular opinion denies the possibility of forcible rape, a judge in England recently suggested that a women who was seriously injured fighting off a rapist had only herself to blame for being hurt. She should have given in to the rapist, said the judge. The *London Times* editorialized, "This almost suggests that refusing to be raped is a kind of contributory negligence" (quoted in *Ms. Magazine*, November, 1977). The accused rapist, a soldier in the Coldstream Guards, was freed pending appeal on the grounds that he has a "promising career"!

In dealing with sexual violence the victim blamers once again ignore the facts. As a whole, according to the National Commission on the Causes and Prevention of Violence, rape victims are responsible for less precipitant behavior than victims of other kinds of crimes (Brownmiller 1976:396). Nor are the vast majority of rape charges brought by "women who changed their mind"; a study showed that only 2 percent of rape complaints proved to be false, which is about the same rate as for other felonies (Brownmiller 1976:410). Finally, the idea that women secretly "enjoy" rape is too preposterous to take seriously. I heartily concur with Herschberger's (1970:24) remark that

> the notion that a victim of sexual aggression is forced into an experience of sensory delight should be relegated to the land where candy grows on trees.

Since the evidence negates the widespread belief that rape victims are "responsible" for what happens to them, it is senseless to argue that if women took special precautions in their dress and behavior the problem would disappear. As Brownmiller (1976:449) convincingly argues: "there can be no private solutions to the problems of rape." Yet these attitudes persist since, by viewing rape as a "woman's problem" brought on by the victims themselves, both men and society are relieved of guilt.

As I suggested earlier, the explanation for and treatment of wife abuse are remarkably similar to that of rape, and the victim blaming is just as loud and clear. Police, who are notoriously loath to intervene in domestic disputes, too often take the attitude "well, if her husband beat her, she probably deserved it." They often assume that women who accuse their husbands of beating them are vindictive, and will only prosecute if they are convinced that the wife is a "worthy victim." And in courtroom after courtroom, it is the battered woman's responsibility to persuade the judge that she is really a victim—a judge who may ask her what she did to provoke her husband's attack.

These attitudes are sometimes shared by members of the abused woman's family as well as society at large. In a newspaper article on wife abuse, a woman whose husband beat her while she was pregnant told of getting no support from her family or her doctor.

> My mother said I must be doing things to make him mad, and my sister said it was all right for a man to beat his wife. I told my gynecologist that my husband was extremely violent and I was mortally afraid of him. Guess what he said? I should relax more. He prescribed tranquilizers. [*Gainesville Sun*, September 5, 1977]

Victim blamers have had a field day in looking for culprits in wife-abuse cases. A member of the New Hampshire Commission on the Status of Women, for example, suggested that the women's liberation movement was responsible for the increased incidence of wife beating and rape (reported on the "Today Show," September 16, 1977).

The fact that a mere two percent of battering husbands are ever prosecuted is clearly related to these attitudes. While assault and battery are quickly punished when they occur between strangers, punitive action is rare within a marital relationship. The extreme to which this can go is evidenced in a recent court decision in England. A man who killed his wife and pleaded guilty to "manslaughter" was sentenced to only three years probation on the grounds that his wife had "nagged him constantly for seventeen years." "I don't think I have ever come across a case where provocation has gone on for so long," said the judge (reported in the *Independent Florida Alligator*, October 20, 1977).

Many who are otherwise sympathetic to the battered wife are perplexed as to why she takes the abuse. But the reasons are not too difficult to discern. Not only are many women economically dependent on their husbands, but they also have been socialized to be victims. As Marjory Fields, a lawyer involved in wife abuse cases, has noted, "they not only take the beatings, they tend to feel responsible for them" (quoted in Gingold 1976:52).

> *Should something go wrong, as in the production of a Hitler, a woman is said to be at the root of the trouble—in this case, Hitler's mother. [Herschberger 1970:16]*

Victim blamers have devoted a good deal of their time and rhetoric to what can be termed "mother blame." In this category of victim blaming, it is not women themselves who are said to be adversely affected by their behavior, but rather their children and, ultimately, society at large. The psychiatric profession, in particular, has been responsible for popularizing the view that, in Chesler's (1972:378) words, "the lack of or superabundance of mother love causes neurotic, criminal . . . and psychopathic children." The absent or uncaring father and other forms of deprivation rarely are blamed for problem children and problem adults.

Mother blame is a natural outgrowth of the traditional, socially approved sexual division of labor that sees child rearing as exclusively "woman's work"; if something goes wrong, it must be mother who is to blame. Moreover, women are held responsible for their children's problems no matter what they do. If they work, they are accused of child neglect, while if they stay home and devote their time to child care, they are berated for smothering their offspring.

The theory that attributes juvenile delinquency and other behavior problems to maternal employment has been around for quite some time. During World War II, working mothers were widely criticized for rearing "latchkey children" who got into trouble for lack of supervision. Mother blame reached a peak shortly after the war with the publication of *Modern Woman: The Lost Sex*. In it, Lundberg and Farnham (1947:304–305) estimate that between 40 and 50 percent of all mothers are "rejecting, over-solicitous, or dominating," and that they produce "the delinquents, the behavior problem children, and some substantial proportion of criminals."

Some years later psychiatrist Abram Kardiner (1954:224) agreed with these sentiments when he wrote that "children reared on a part-time basis will show the effects of such care in the distortions of personality that inevitably result." After all, "motherhood is a full-time job."

Lest it be thought that mother blame is merely an artifact of the days of the feminine mystique, a recent newspaper editorial espoused it when attempting to explain the high crime rate.

> Let's speculate that the workaday grind makes Mom more inaccessible, irritable . . . and spiteful, thereby rendering family life less pleasant . . . than the good old days when she stayed in the kitchen and baked apple pies. What could that be doing to the rising crime rate? Say fellows, could it be Mom's fault? [Editorial, *Gainesville Sun*, March 25, 1977]

These views, of course, ignore the studies that indicate that absent and low profile fathers are more responsible for delinquency in their children than are working mothers. Moreover, the most comprehensive study of maternal employment, *The Employed Mother in America* (Nye and Hoffman 1963), effectively rebuts the myths concerning the supposed ill effects of working mothers on their offspring, and concludes that "maternal employment . . . is not the overwhelming influential factor in children's lives that some have thought it to be" (Burchinal 1963:118; Siegal et al. 1963:80). But, as we have seen, victim blamers have little use for facts that contradict their strongly held beliefs.

What of the woman who stays home and devotes full time to child raising? She too is the target of the mother blamers who hold her accountable for an incredible variety of social problems. In his book *Generation of Vipers*, Phillip Wylie (1955) characterized such women as "Moms" who led empty lives and preyed upon their offspring, keeping them tied to their proverbial apron strings. This theme was also sounded by Edward Strecker (1946), a psychiatric consultant to the Army and Navy Surgeons General during World War II. In trying to account for the emotional disorders of 600,000 men unable to continue in military service, Strecker wrote, "in the vast majority of case histories, a Mom is at fault." But what causes "Moms" to be the way they are in the first place? In most cases, a Mom is a Mom because she is the immature result of a Mom, says Strecker (1946:23, 70).

> *If it weren't for Martha, there'd have been no Watergate. [Richard Nixon on David Frost Interview, September, 1977]*

Richard Nixon's statement holding Martha Mitchell responsible for Watergate is a timely reminder of the length to which victim blamers sometimes go. According to Nixon, Watergate occurred because "John Mitchell wasn't minding the store," but was preoccupied with his wife's emotional problems. This claim is particularly malicious given the often-noted fact that the large Watergate cast was *all* male. A similarly absurd remark was made during the "Son of Sam" episode, when it was automatically assumed that a woman was at the root of "Sam's" problem. The killer "must have been terribly provoked by a woman," New York psychiatrist Hyman Spotnitz was quoted as saying in *Time* (July 11, 1977).

While Nixon's assertion was widely seen as self-serving, the opinions of psychiatrists and other authoritative victim blamers are taken quite seriously by the general public, including women. Women not only participate in this ideology, they often internalize it, blaming themselves and other women for a host of problems. This shows the effectiveness of the ideology in rationalizing subordination to the victims themselves. The very persistence of victim blaming, in fact, is partly due to the implicit participation of its targets. And men, of course, perpetuate the ideology since it is clearly in their own self-interest to do so. It helps maintain the status quo from which they benefit.

In recent years the ability of victim blaming to deflect attention from social institutions and obscure societal processes has been particularly valuable in "explaining" women's failure to make significant advances in employment and other realms. Despite the existence of the feminist movement and a plethora of equal opportunity laws, women still overwhelmingly remain in low-paid, low-prestige, female job ghettos. But, say the victim blamers, that is because they have no interest in getting ahead, they fear success, and don't want the added responsibility that comes with promotions. The goal of the victim blamers is clear: these purported qualities of the victimized group conveniently mask the fact of continued widespread sexual discrimination. But it must be emphasized that although the Blaming the Victim ideology does distort reality by covering up the inequalities in contemporary American life, it is not the *cause* of these deeply rooted social and economic inequalities: it is a rationalization for them.

This content analysis of eight "women's magazines" complements Margolis's observations about treatment of women in the contemporary mass media. Patricia Rentz's research project was a quantitative study of stories and advertisements in *Teen, Seventeen, Glamour, Cosmopolitan, Mademoiselle, Ms., Redbook,* and *Family Circle.* She finds that these magazines present an inconsistent and contradictory view of appropriate female roles. With the exception of *Ms.* magazine, where the feminist perspective is consistent, advertisements in women's journals portray women in traditional roles, while stories stress modern roles, problems, and opportunities. Like other selections in this volume, Rentz contributes to the study of tacit enculturation, arguing that by tacitly enculturating women into traditional roles through advertising and into modern roles through stories, magazines send women contradictory messages about the roles they should and do perform. Furthermore, this contradiction confronts women with the cultural desirability of succeeding in both roles, an unrealistically demanding goal.

22 / **A Picture Is Worth a Thousand Words**

Patricia M. Rentz

In the United States, social scientists as well as laymen are concerned with the roles men and women perform. Anthropologists, sociologists, and psychologists study the basis for these acquired roles, how the sexes attained such seemingly diametric positions, and the effect these roles have had on people in the past as well as the effect on men and women in contemporary society. With the onset of the feminist movement in the late 1960s, sex roles have gradually changed. The goal of the feminist movement is not to create a one-sexed society, but to achieve and increase equality between the sexes.

> In no way do we want to become men. We are women and proud of being women. What we do want to do is reclaim the human qualities culturally labelled "male" and integrate them with the human qualities that have been seen as "female" so that we can all be fuller human people. [Boston Women's Health Book Collective 1976:18]

Already, progress toward this goal may be noted in that many businesses and colleges claim to be equal opportunity employers. The University of Michigan, for example, upholds that it is "an equal opportunity/affirmative action" employer. Under applicable federal and state laws, including Title IX of the Education Amendments of 1972, the university is not to discriminate on the basis of sex, race, or other prohibited matters in employment, in educational programs and activities, or in admissions (Reidel 1978). Thus, to amend previous practices, the university, as well as many other businesses, colleges, and organizations, no longer discriminates on the basis of sex or race. Although discriminative practices and definitions of sex roles are continually changing for the better, much progress is still necessary to create sexual equality.

In an attempt to discover the roles women play in contemporary society, magazines will be studied. Noting the strong impact the mass media have on influencing and maintaining sex roles, it will be interesting to discover if and how the image of modern women portrayed by the mass media correlates with the modern image of women held by feminists—that of women as competent, freethinking, and independent persons. To achieve equality between the sexes, some of the mass media are striving to replace the image of the "traditional" woman with the image of the "modern" woman. It is hypothesized that though many magazines claim to be directed at the modern American woman and her new image of being strong, independent, and freethinking, they fall short of this desired goal in that their advertising disclaims and disagrees with the content of the magazine. Articles may illustrate this modern notion of the strong, independent, and assertive woman, but the advertising does not reinforce these ideas, since it portrays women in the traditional roles of being weak, passive, and dependent. Thus, because of these contradictory messages, many women are confused as to what role they perform in contemporary society.

To test this hypothesis, only women's magazines will be studied. To include other magazines that do not deal primarily with women would mean bringing in irrelevant data. If articles not focused on women were also studied, the hypothesis could not be tested directly to determine if the magazine is giving contradictory messages of women being assertive, strong, and independent in stories, and passive, weak, and dependent in advertising. In a broad range of magazines read by American women—*Teen*, *Seventeen*, *Glamour*, *Cosmopolitan*, *Mademoiselle*, *Ms.*, *Redbook*, and *Family Circle*—content and advertising will be studied in an attempt to prove the hypothesis that magazines present an inconsistent picture of the modern American woman.

Two magazines read primarily by female teenagers, *Teen* and *Seventeen*, were selected in an effort to study the portrayal of women to teenagers. To determine the image of women presented to females in their late teens and twenties, *Glamour*, *Cosmopolitan*, and *Mademoiselle* were selected. Two magazines, *Family Circle* and *Redbook*, were selected as appealing primarily to "older" women. *Ms*. magazine was chosen since it appeals primarily to feminists. Thus, in this broad range of magazines, appealing to many different women, a clear picture of the mass media's idea of modern American women can be studied.

Two issues of each magazine were studied by content and advertising. The products advertised and the role women portray in each advertisement were investigated. To obtain a consistent data set, only the products most commonly advertised in all magazines were studied. These advertisements include: business/career choices, clothes, cosmetics/beauty aids, cigarettes, home care/food. "Clothes" includes shoes, clothes, and underwear; "cosmetics/beauty aids" includes shampoo, deodorant, and makeup; "home care/food" includes washing detergents, pet food, and money-saving coupons. Each advertisement was then judged to determine if it depicted women in traditional (T), modern (M), or dual roles (D). Traditional roles present seemingly unintelligent, dependent, nonassertive, weak, and passive women. Such a woman is usually a gorgeous blonde with a very revealing dress on; more often than not, she has a handsome man by her side. If she is not a gorgeous blonde, she is probably a beautiful housewife with kids surrounding her, looking bewildered, with a man giving her instructions. "Modern" women portray the new image of being strong, independent, and assertive. These women, while still beautiful, are usually pictured alone in atypical occupations or practices. Dual roles are defined as those women depicted in nontraditional roles while maintaining the dress and looks of the traditional female. More often than not, a male is by her side, helping and instructing her. After studying the advertising, the content of each magazine was studied to determine if the stories attempted to portray women in modern roles, as assumed in the hypothesis. The story content in each magazine was surveyed to determine which types of stories were most common among all the magazines. The categories used in judging contents were: child care, sports, cooking, business, college, female awareness, job hunting, pregnancy/abortion/marriage/divorce/sex, and money management. To clarify some of these ambiguous headings, "female awareness" includes articles on feminism and self-awareness/improvements (i.e., handling guilt, being more assertive, handling stress), and the category entitled "pregnancy/abortion/marriage/divorce/sex" is so large because I found that many of these topics were handled simultaneously within the same article. For each category, the roles women portrayed were judged to be either traditional, modern, or dual. The "traditional" stories depicted women as weak, passive, and dependent, while the "modern" stories portrayed them as strong, assertive, and independent. The "dual" stories depicted women as being able to have outside jobs, while still "naturally" holding the traditional jobs of mother, cook, cleaner, and seamstress. Thus, in judging advertising and content by the criteria and categories previously mentioned, the hypothesis can be tested. Advertising will be studied to determine if in fact women are portrayed in traditional roles, while content will be studied to determine if in fact women are portrayed in modern roles. If these two assumptions are upheld, it may be stated that magazines give two contradictory messages to women, thus possibly confusing them about their proper role in contemporary society.

The compiled data for two issues of each magazine are presented in table 1. Only the categories of advertising and content that were applicable to each magazine are mentioned. For the rest of the categories, the reader may assume that no such advertisements or stories were found in the magazine.

In interpreting these data, each magazine will be dealt with separately; afterward, an overview of the combined data will be provided in order to see if any patterns and correlations are apparent.

In viewing *Teen*, a magazine focusing on beauty tips and advice to young teenage girls, conclusions may be drawn regarding the role women hold in society by studying advertising and content. As cosmetics were the only products advertised, and the vast majority of women portrayed held traditional roles, one may conclude that teenage girls are being shown visually that the contemporary role of women is still the traditional role. On the other hand, women were portrayed in modern roles in stories three times as often as in traditional roles. For instance, in an article under the category of female awareness entitled "Women's Lib—What Guys Think" (November, 1977, p. 22), the author wrote that 99 percent of the young men interviewed believed in equal rights for women and gave reasons for their beliefs. Thus, with *Teen* magazine portraying women in traditional roles in advertising in a 30T:7M:1D ratio, and modern roles in content in a 3M:1T ratio, the hypothesis is upheld. By reinforcing the traditional role of women in advertising, and introducing or reinforcing the modern role of women in content, young women are given a contradictory picture of the role they play in contemporary society.

For *Seventeen*, a magazine focusing on advice and stories about

TABLE 1 Portrayal of Women's Roles and Related Topics in Current Women's Magazines

Category	*Teen*	*Seventeen*	*Glamour*	*Cosmopolitan*	*Mademoiselle*	*Ms.*	*Redbook*	*Family Circle*
ADVERTISING								
Business		3M[a]	5M	3M		1M	1D	1M
Cigarettes			3T,6M	7T,6M,1D	3T,4M	5M	11T,6M	3T,6M
Clothes		26T,8M,2D	45T,1M,1D	42T,13M,3D	27T,4M,2D	1T,1M	16T	1T
Cosmetics	30T,7M,1D	61T,10M,3D	107T,1M,1D	107T,10M,2D	89T,3M	3T,2M	47T,1D	13T
Home care		1T		6T,2M	1M		13T	12T,2M
Total	30T,7M,1D	88T,21M,5D	155T,17M,2D	162T,34M,6D	119T,12M,2D	4T,9M	87T,6M,2D	29T,9M
CONTENTS								
Business				1M	1M	3M		
Child Care			4M	1M				
College		3M	1M		2M			1M
Cooking	2T			1M				1M
Female awareness	2M	5M	8M	7M	1T,1M,1D	6M	3M,1D	1T,2M
Job hunting	1M	2M	1T,5M	1M	1M		2M	
Money management		1D	8M	3M	1M		1M	1T,1M
p/a/m/d/s[b]	1M		8M	11M	3M	5M	6M	1M
Sports	2M	7M	5M	1M	1M	2M		
Total	2T,6M	17M,1D	1T,39M	26M	1T,10M,1D	16M	12M,1D	2T,6M

a. T = traditional roles; M = modern roles; D = dual roles

b. p/a/m/d/s = pregnancy/abortion/marriage/divorce/sex

matters that concern the female in her midteen years (i.e., college, job hunting, sports), conclusions are similar to those drawn from *Teen* magazine. Since more products than cosmetics were advertised, women were seen in a variety of advertisements, but still in traditional roles most of the time. An example of a modern role is an advertisement for blue jeans with a female pilot in an airplane. An example of a dual role portrayed a woman in lacy underwear (traditional) with running shoes on (modern). In stories, women were always seen in modern roles, except for one dual role. The dual role instructed young women how to manage money more efficiently, but they were told to do so in order to buy more clothes. Thus, with *Seventeen* magazine portraying women in traditional roles in advertising in a 88T:21M:5D ratio, and in modern roles in content in a 17M:1D ratio, the hypothesis is upheld. *Seventeen* reinforces the traditional notion of women in advertising, while it introduces or reinforces the modern role of women in stories, thus presenting a contradictory picture of the role women hold in contemporary society to females in their midteen years.

Glamour is a magazine devoted to keeping the woman in her late teens and early twenties up-to-date on fashion trends and on other aspects of her life (i.e., career, college, etc.). Judging from the results of the survey, which reveals that women are portrayed in traditional roles in advertising in a 155T:17M:2D ratio, and in modern roles in contents in a 39M:1T ratio, the hypothesis is once again upheld. Through these contradictory messages, where women see one role model and read about another, some women may be confused as to what role they hold in contemporary society.

Cosmopolitan, a magazine read primarily by women in their twenties, repeats the findings of the former magazines. Since women are portrayed in advertising in traditional roles in a 162T:34M:6D ratio, and depicted in modern roles in contents in all twenty-six stories, the hypothesis is upheld.

Mademoiselle is read primarily by women in their twenties and focuses on informing and advising women on the matters that interest them (i.e., career, job hunting, money management), as well as keeping them up to date on the season's latest fashions. The hypothesis is proven true once again, as women are illustrated in traditional roles in advertising in a 119T:12M:2D ratio, and depicted in modern roles in content in a 10M:1T:1D ratio.

Ms. magazine appeals primarily to feminists and is read by women of all ages. It deals with feminist issues in the United States as well as with feminists in other parts of the world. *Ms.* magazine *refutes* the hypothesis, since women are consistently shown in modern roles in both advertising and content.

Redbook appeals primarily to women in their late twenties and older. It focuses on giving ideas and advice to women concerning home, health, careers, and "self-improvement." As women are portrayed in traditional roles in advertising in a 87T:6M:2D ratio, and depicted in modern roles in content in a 12M:1D ratio, the hypothesis is upheld once again.

Family Circle, as the name implies, is read primarily by women with families. Judging from the results of the survey, the hypothesis is upheld, as women are portrayed in traditional roles in advertising in a 29T:9M ratio, and portrayed in modern roles in content in a 3M:1T ratio.

Thus, in evaluating the data overall, the hypothesis is confirmed in seven of the eight magazines surveyed: though many magazines claim to be directed at the modern American woman and her new image of being strong, independent, and freethinking, they fall short of this goal since their advertising disclaims and disagrees with their stories. Articles may illustrate this modern notion of women, but advertising portrays women in the traditional role of being weak, passive, and dependent. Magazine advertisements portrayed women in traditional roles most of the time to sell, for example, shampoo promising silkier, sexier hair or cigarettes telling the woman "you've come a long way, baby." (It's amazing she's done anything other than regress with a slogan like that!) Some advertisements portrayed women in dual roles, such as a businesswoman acquiring a job through a placement agency entitled "Kelly Girls." (If "girls" aren't women by the time they are in business, when do they ever become women?)

While advertisements depict women in traditional roles, it is heartening to note that women are portrayed in modern roles in the majority of stories. Through such articles as "Women—We Have Influence" (by Rosalynn Carter, *Redbook*, September, 1979, p. 110); "What if His Dreams Don't Match Your Dreams?" (*Redbook*, August, 1979, p. 39); "Dollars and Sense, What You Should Know About Your Job Rights" (*Mademoiselle*, July, 1979, p. 164); and "A Chance at a Second Life—Facts about Widows" (*Family Circle*, April 3, 1979, p. 14), women may read and understand their new place in society. Traditional and dual roles were also present in articles such as "Twelve Ways to Save on the Basics" (*Family Circle*, September 18, 1979, p. 86), which has illustrations of women in all the traditional roles of shopper, cook, and cleaner; and "Homeworks—Ten Ways to be a Great MANAGER" (*Redbook*, August, 1979, p. 69), which portrays dual roles by discussing homemaking in business terms. One must understand that not all traditional stereotypes can be broken in advertising and story lines. This approach would make the notion of the modern woman too unrealistic for some women to believe. *Ms*. magazine can depict women in modern roles in both advertising and content because it appeals to feminists. These women fully understand their modern role in contemporary society and expect to see it illustrated consistently throughout the magazine. In fact, toward the end of the magazine, there is a regular feature entitled "No Comment" in which readers submit copies of sexist advertising.

The implications of this study are that women are in fact given contradictory messages about the roles they play in contemporary society. Young women who are internalizing role behavior may be confused when they read *Teen* and *Seventeen* magazines. They read that women are strong, independent, and assertive in stories, and yet they see women who are weak, passive, and dependent in advertising. Unless these young women already have strong ideas of their role as either traditional or modern, they are likely to be confused when they read the magazines. Women who are in their twenties reading *Glamour*, *Mademoiselle*, and *Cosmopolitan* usually have a fairly firm notion of their role in contemporary society. They may either identify with the traditional women in advertisements and not identify with the modern women portrayed in the magazines' articles; or they may view advertisements for the products only and ignore the role of women in them, and identify with the modern notion of women presented in stories. Older women reading *Redbook* and *Family Circle* probably have their notion of the role they play in society fixed. If they value the modern role of women, they are satisfied with the material they read. If they play the traditional role, articles may still be valuable. Those "traditional" women may even be inspired to change their roles as they see other women their age doing the same, (e.g., in "Back to College at 34," *Family Circle*, September, 1979, p. 58). Thus, through the interpretation of the data, it is implied that those contradictory messages have a different impact on each audience.

Through the explanation and testing of the hypothesis, one may fairly surmise that women's magazines present inconsistent, contradictory messages to their readers. The main emphasis now is what these contradictory messages reveal about contemporary American culture. As Alan Graebner (1975:23) reveals in his article "Growing Up Female," "the transmission of the tacit culture is subtle, but its content is basic." This transmitting of the proper sex roles by magazines is tacit, as it is done subliminally. Magazines must satisfy readers to stay in business, so they portray women in modern roles in stories. An area in which readers do not have as much control is advertising. Unfortunately, the products that illustrate women in traditional roles are usually food and clothes, products that most women could not do without (except for non-essential cosmetics, which also usually portray women in traditional roles). Thus, by illustrating women in traditional roles in advertising and modern roles in stories, magazines (and more generally, the culture) seem to be telling American women that they must maintain their traditional roles even as they acquire their modern roles. That is, contemporary society expects a woman to *inwardly* maintain her features of being weak, passive, and dependent even as she assumes the characteristics of being more assertive, independent and freethinking *outwardly*. Thus, by tacitly enculturating women into traditional roles through advertising, and enculturating women into modern roles through stories, females are given contradictory messages about the roles they play, and society expects them to adopt both of them.

Also, this study could be interpreted in the sense that it may reveal something about the modern American woman. A case could be made that she realizes what her modern role is supposed to be, but she may not be emotionally prepared to deal with it, and prefers the comfort of the traditional role. This point however, while having merit, should be the topic of another research project.

Patricia Hesseltine's analysis of the content of 389 television commercials, a sample that covered every minute of air time between 6:00 A.M. and 2:00 A.M., echoes many of Rentz's findings (this volume) about tacit sex-role enculturation in the contemporary United States. However, whereas at least in women's magazines the stories offer the possibility of nontraditional female roles, women in television commercials are almost invariably portrayed as wives, mothers, and homebodies. Despite the fact that more than 42 percent of the American work force consists of women, Hesseltine shows that of 108 commercials in which a woman's occupation could be determined, 83 percent were home-oriented; only eleven occupations other than housespouse were shown. Hesseltine also found that women in television commercials are frequently portrayed as stupid and incompetent, especially in their roles of home manager, wife, and mother. Cats know what they want (and get it) more often than women. Hesseltine's detailed analysis of timing and content of TV commercials demonstrates a shocking lag between fairly widespread feminist consciousness and the tacit lessons offered by the media.

23 / **The 1980 Lady as Depicted in TV Commercials**

Patricia Hesseltine

In this essay I will examine tacit enculturation through television advertising. Tacit enculturation is a mysterious phenomenon. Through tacit enculturation, values and attitudes are transmitted silently. Tacit enculturation occurs without conscious verbal instruction and in many cases the "teacher" does not know a transmission is taking place.

Much of our socialization is tacit, yet the learner does not know that critical information is being transferred, and that it is molding his or her character for a lifetime. Tone of voice, subtle body movements, facial expression, spatial arrangements of objects and people, unspoken (often unexamined) expectations, things left unsaid, are all part of the tacit dimension of enculturation (cf. Graebner 1975).

Television enculturates both explicitly and tacitly. Almost every household has at least one of these machines drawing the attention of its inhabitants for a good many waking hours. Culture comes careening into the living rooms of millions of Americans, influencing values and attitudes from dawn until 3:00 A.M. What is this medium telling Americans about roles, attitudes, and values with respect to females?

The nature of television programming is important; however, the corporations that finance the specific programs control the air waves and have much to say about the content of the programs. For this reason I have decided to investigate the nature of commercial advertising messages to understand the types of attitudes Americans will accept in commercial messages.

Corporations must project in fifteen to sixty seconds a micromessage with maximum salespower that will appeal to the majority of the American public. The television shows themselves have about a half hour or an hour to accomplish the same objective. Both the shows and the advertisements must contain values and attitudes that are completely American, or they will not stay on the air, or their products will not sell.

For this study I watched television for several days until I had gathered data from every minute of air time, that is, for the entire width and breadth of the telecasting day, from 6:00 A.M. until 2:00 A.M. I wrote down every commercial, its contents, who were major actors, what their roles were in the advertisement, who became the authority figure, and, if possible, or if shown, what the actors' occupations happened to be. During this twenty-hour period 389 commercials were recorded.

I began with the basic question "How does television define women?" After viewing twenty hours of TV and the 389 commercials, the answer was clear. Men were the experts on all grocery items, television and stereo sets, advanced technology, furniture, carpets, law, medicine, restaurants, banking, and the stock market, to name a few. Women knew about tampons, fake fingernails, dog food, and, in some cases, cough medicine and panty hose.

I divided my data into products that appealed to both sexes (363 commercials) and those that appealed to only one sex. I lumped together all grocery and paper products into the category that appealed to both sexes; however, I did so with mixed emotions. Television sets, appliances, and furniture also fell into this joint category. In the cases where the products appealed to both sexes, 76 percent of the time males were the experts; 7 percent of the time both sexes had equal input; 13 percent of the time the female had the final word; and 4 percent of the time (15 cases) either children or pets knew what was right for them and became the authority.

It is interesting to note that cats were projected as total experts on what they wanted and needed 66 percent of the time, whereas women knew only 13 percent of the time what they wanted when the product appealed to both sexes. Moreover, children were the experts in two out of every three commercials they appeared in. It was also interesting to see that most of the children were boys—boys who demanded that their mother produce certain goods.

For example, one hot dog commercial shows the father and son unilaterally rejecting the cheap, inferior hot dog the mother has purchased to save money, in favor of the advertised product. The mother looks distressed and upset because her cooking has been spurned, and this is only rectified when the demands of father and son are met with these particular hot dogs. The family is once again a cohesive unit enjoying their favorite brand of hot dogs, which the mother had to go out and buy—theoretically throwing out the rejected brand and her cooking.

There is also a catsup advertisement in which Shorty, his mother, and Shorty's favorite catsup are introduced. Shorty's proud mother serves Shorty's favorite snack with the catsup that Shorty enjoys the most. Shorty is a six-foot-three-inch bruiser of a child who looks as if he could easily break his tiny mother in two if she chose the wrong product. Force is not overtly suggested, but tacitly implied, by sexual dimorphism.

Little boys appeared in cereal, gelatin, milk, candy bar, cough medicine, potato chip, and cookie ads and were experts. Little girls didn't know much about anything, and if they did appear they were never demanding personalities who asserted their needs and wants like the boys. They appeared as traditional nonassertive females. (This is certainly an interesting beginning for the "liberated females" of tomorrow.)

When children reached their teenage years they were always shown in couples. These couples interacted on all occasions, but generally the female was a complainer, and the male the expert about the product that would make her happy. This was particularly true when it was a product that both sexes could use or consume. However, in one advertisement a young girl brings home her boyfriend after a game and announces to her mother that "Randy (the boyfriend) was the star and saved the game." The humble boy looks hungry after all that activity, and the mother offers to make him a sandwich. However, the teenage girl knows better than her mother . . . "Randy would prefer a Manwich." The mother, by facial expression, concedes that her daughter knows best and sets out to prepare the Manwich.

During my twenty hours of recording commercials, I saw only one ad that appealed only to men. The ad was for a men's shoe store and the commercial was done by a man. However, in eleven of the twenty-five commercials geared exclusively to women, a male was the final authority on the product (see table 1). This means that 44 percent of the time a male was an expert on a product used exclusively by women. These included beauty salons, hair-care products, bath powder, lipstick, deodorants, women's hair conditioners, skin lotions, and panty hose. However, the males did not corner the authority market on panty hose; females were experts 50 percent of the time in this area (one out of two commercials). Of the remaining 56 percent of the commercials geared only to women (fourteen out of twenty-five), 57 percent of the time these commercials were for tampons, sanitary napkins, or douche products (eight out of fourteen). This area was totally taboo to men, and women excelled as experts.

TABLE 1 Commercials That Appealed Only to Females

		Authority	
Time	Product	Male	Female
7:25 A.M.	Fashion Beauty Salon	x	
7:40	Curlers	x	
7:50	Panty hose		x
8:30	Beauty Salon		x
8:35	Sassooning (hair conditioner)	x	
8:40	Beauty salon	x	
9:40	Bath powder	x	
10:30	Rely (tampons)		x
11:10	Fancy Fingers (artificial fingernails)		x
11:30	Lipstick	x	
11:45	Women's deodorant	x	
12:00 P.M.	Hair conditioner		x
12:25	Fancy Fingers		x
1:10	Sassooning (hair conditioner)	x	
1:15	Mini-pads		x
1:25	Douche		x
1:25	Hair spray		x
3:10	Mini-pads		x
3:15	Mini-pads		x
3:50	Panty liners		x
4:00	Porcelana	x	
	Nothing for four hours		
9:00	Oil of Olay	x	
11:00	Rely (tampons)		x
12:15 A.M.	Panty hose	x	
1:15	Panty liners		x
Total		11	14

The uninformed housewife scenario occurred over and over. Teenagers, small children, husbands, older women, and foreign women took turns telling the ignorant housewife what to purchase.

Women were cast as housewives and mothers in most of the ads in which they appeared; however, as an occupational group, they seemed to know little about their occupations. Although their main role was that of caretaker or serving person, their stupidity arose throughout the viewing day and evening, but was *most prominent* during hours when children would watch, and, interestingly enough, just before adults' bedtime. Possibly these commercials, which depict women as stupid, are shown just before bed because advertisers hope that women will go to bed carrying the suggestion that they are stupid and "clean up their acts" while shopping the next time.

The other time period when women are depicted as stupid falls within the children's viewing hours. One woman cannot tell dog food from stew meat, and this commercial is shown over and over. It is interesting that these "stupid mother and housewife" commercials are shown at times when professional, working people cannot see them—too late or early in the day.

In table 2, I list the women's roles and occupations (where implied), as they were presented throughout the entire telecasting day. This table represents 108 women in commercials in which an occupation or role was implied. If this is an accurate sample of the images being projected by the media in the advertising commercials, 83 percent of women's occupations or roles are home-oriented. In this sample only eleven occupations other than housewife were shown. Of these eleven occupations, seven were job slots traditionally held by women: secretary, nurse, teacher, manicurist, hospital therapist, waitress, and airline hostess. Of the remaining four occupations—computer technician, skin diver, corporation owner, and potential auto mechanic—two came as part of General Electric's sponsorship of a two-hour movie. This high-powered advertising was geared to the "now generation" and was exceptional among the commercials I saw. Forty-two percent of the traditional occupations were also contained in the General Electric advertising. The potential female auto mechanic was joining the Army and was seen at 1:00 A.M. This leaves the construction firm owner as the only nontraditional female for children to identify with during the time they usually watch television. This woman was an obese, matronly person with no charisma. The occupations depicted during the hours children would be watching were housewife, mother, nurturer, man's companion, stupid woman, manicurist, waitress, and secretary. This is a powerful source of tacit enculturation and perpetuates the status quo.

Tacit enculturation about occupational and role models, then, is telling young and old males and females alike that women should remain in the home and that their role and occupation should be that of the traditional housewife. Americans think in terms of binary oppositions, and television commercials verified this again and again, as stupid mothers or wives were helped to overcome their ignorance about their roles by men. Women were generally depicted as middle-class people in middle-class kitchens, whereas men were always providers and protectors who, when depicted alone, were distinguished, well-dressed, and informed. When they interacted with women, they were usually authorities about whatever product they represented, unless they were sick.

The coffee commercials became a source of enjoyment and confirmed the findings of Lawrence Taylor (1976), who states that coffee symbolizes, among other things, the household itself. Without exception coffee ads depicted men as superior. One advertisement showed a pompous Hispanic man touting the wonders of coffee and stated that "Coffee should have character like a man." Not one of the 389 commercials assigned a parallel characteristic to a woman. Women were advised how to please men and friends, and were depicted as the males' companions in the remainder of the coffee commercials. In one commercial both the mother and the father are sitting in an elegant room watching their daughter play the cello. This seems to me to be the essence of family stability.

The zippiest ad involving a woman was the "1980 lady" commercial for Pillsbury cake mix, which probably contained the best and the worst of the tacit enculturation. It was implied that this woman was something other than a housewife, caretaker, wife, mother, nurturer, and sex partner, but the bottom line was that *even though* she might be something else, she baked like an old-fashioned girl.

Many other American values were presented in tacit form in the commercials: protection of private property through insurance; voluntary group association, such as Weight Watchers; materialism through mass consumption of such luxury goods as TVs, autos, vacations; clock and calendar mentality, as people were constantly in a hurry, looking at their watches, but being served instantly. The values of individual achievement and upward mobility could be seen in the status of men in the commercials and the mythical classlessness of American culture showed up in the uniform portrayal of all the women as middle class.

TABLE 2 Women's Roles or Occupations as Presented in Commercials

Time	Role or Occupation	Product
6:45 A.M.	Sage elder	English muffins
7:40	Housewife/caretaker	Cold medicine
7:50	TV star (Barbara Eden)	Panty hose
8:00	Housewife	Granola bars
8:00	New mother	Diaper service
8:30	Housewife/mother	Aspirin
8:50	New mother	Diaper service
8:50	Mother	Soup
8:58	Nurturer	Dog food
9:00	Owner of home improvement company	Home construction
9:40	Carpet expert (local female TV star)	Carpet
10:20	Wife	Frozen fish
10:30	Male's companion	Packaged cheese
10:35	Housewife	Dish soap
10:35	Mother	Chicken mix
10:35	Housewife	Washing soap
10:40	Good mom	Powdered soft drink
10:50	Housewife	Laundry product
11:00	Frustrated mother	Paper towels
11:10	Housewife/cook	Vegetable oil
11:25	Mother	Dish soap
11:30	Sex symbol	Lipstick
11:30	Wife	Deodorant
11:30	New mother	Diaper service
11:45	Caretaker of sick husband	Cough medicine
11:45	Nurturer	Cake mix
11:55	Airline hostess	Headache remedy
11:55	Famous actress	Tissue
12:15 P.M.	Mother/caretaker	Cold medicine
12:30	Cook	Soup
12:45	New mother	Baby soap
12:50	Stupid mother	Chips
12:50	Wife	Frozen fish
1:00	Wife	Vegetable oil
1:20	Frustrated housewife	Paper towels
1:20	Housewife	Vegetables
1:25	Perfect for job/none suggested	Hair spray
1:30	Mother	Packaged cake products
1:30	New mother	Diaper service
1:30	Housewife	Paper towels
1:45	TV star (Joann Worley)	Tissue
1:45	Housewife/shopper	Room deodorizer

TABLE 2—*Continued*

Time	Role or Occupation	Product
1:50	Nurturer	Cake mix
1:50	Companion	Coffee
2:00	Good mom	Powdered soft drink
2:00	Old-fashioned woman	Syrup
2:07	Caretaker mother	Cold medicine
2:15	Stupid wife	Dog food
2:30	Mother/caretaker	Diapers
2:45	Mother/caretaker	Laundry soap
2:45	1980 lady	Cake mix
2:50	Mother/caretaker	Cold medicine
3:00	Nurturer	Cookie mix
3:10	Caretaker/sick husband	Cold medicine
3:10	Stupid shopper	Dog food
3:15	Manicurist	Dish soap
3:15	Wife	Deodorant
3:30	Cook/nurturer	Cookies
3:30	Housewife/cleaning person	Cleaning liquid
3:35	Cook/nurturer	Soup
3:35	Female corporation owner	Home construction
3:45	Uninformed mother	Canned meat spread
3:50	Companion	Frozen fried potatoes
3:50	Parent/proud mother/wife	Coffee
4:00	Caretaker	Cough medicine
4:12	Cook/nurturer	Soup
4:40	Laundry person	Laundry soap
4:40	Stupid mother	Hot dogs
4:55	Waitress	Seafood restaurant
5:40	Waitress	Beer
Noticeable absence of women from ads during this time frame.		
6:35	Housewife	Coffee
6:35	Dumb pet owner	Dog food
6:35	Stupid mother	Hot dogs
7:10	Helpful wife/caretaker	Cough medicine
7:15	TV star (Marion Ross)	Laundry product
7:20	Caretaker	Aspirin
7:30	Cook/nurturer	Spaghetti
7:30	Companion	Coffee
8:15	Old-fashioned woman	Syrup
8:15	Companion	Coffee
8:30	Concerned mother	Toothpaste
8:30	Secretary	Vitamins
8:30	Favorite mom	Powdered soft drink
9:00	Mother	Telephone company
9:00	Secretary	Appliance manufacturer

TABLE—*Continued*

Time	Role or Occupation	Product
9:00	Housewife/shopper	Coffee
9:15	Skin diver/researcher	Appliance manufacturer
9:15	Nurse	Appliance manufacturer
9:45	Computer technology/1 female, 6 males	Appliance manufacturer
10:25	Hospital technician	Appliance manufacturer
10:30	Handicapped housewife	Appliance manufacturer
10:45	Nurturer	Dog food
11:45	Stupid wife	TV sales room
11:45	Secretary	Cold tablets
12:00 A.M.	TV star (Joann Worley)	Tissue
12:00	Cook/nurturer	Appliances
12:15	Dependent wife	Insurance
12:15	Housewife	Dish soap
12:30	Mother	Catsup
12:30	Caretaker/sick husband	Cough drops
12:45	Stupid shopper	Dog food
12:45	Frenzied woman/housewife	Water softener
12:45	Nurturer	Dog food
1:00	Working woman/nondescript occupation	Meat mix
1:00	Stupid mother	Chips
1:00	Teacher	Digestion mint
1:00	Future mechanic	U.S. Army
1:15	Manicurist	Dish soap

The wide variety of male occupations shown (chef, carpet expert, furniture expert, stock broker, famous hair stylist, upholsterer, fabric expert, legal sage, magazine expert, pilot, detective, air traffic controller, tax consultant, engineer, technologist, inventor, highway safety expert, and TV repairman) contrasts sharply with the limited range of female occupations. Although the male could easily hop into the kitchen or the laundry room, the female did not jump out of the kitchen into the pilot's seat or onto the New York Stock Exchange.

Thus, in today's television commercials, even in a time of female liberation, women still cannot extricate themselves from the image of the kitchen and laundry room. This stereotype is being taught every day, through tacit enculturation, to the youngsters of America, as well as to its frustrated housewives.

Christina Hill provides a detailed analysis of the portrayal of blacks, their interrelationships, and their relationships with whites in daytime television soap operas through fall, 1979. She compares some of the social attributes of these television blacks with the actual social conditions that affect black Americans. Hill focuses on residential segregation, unemployment, occupation, upward mobility, and dialect to show that the soaps portray blacks unrealistically. Among the misrepresentations: television blacks are not residentially segregated; are always employed; have high-status, high-income occupations; are socially mobile; almost always speak the standard dialect; have undeveloped romantic relationships; and have little black ethnic consciousness. Hill suggests that by depicting blacks as more successful than they are in reality, soaps block most Americans' understanding of the problems that do affect blacks and the remedies that are necessary to solve them. Success is portrayed in terms of individual achievement rather than in the context of social conditions that affect whole groups of people. Here we have another example of the Blaming the Victim ideology, a fundamental and pervasive belief of American culture.

24 / **Blacks on Daytime Television**

Christina D. Hill

Introduction and Statement of Hypothesis

Daily, millions of Americans tune in and "turn on" to television soap operas. Soaps have become an important part of many Americans' lives. In recent years, television serials have expanded the roles of blacks. But do soap operas portray blacks realistically, or are they merely white characters with darker pigmentation? My hypothesis is as follows: in terms of housing, segregation, and occupational status blacks are unrealistically portrayed on television serials. I will examine whether the position of black soap opera characters reflects the position of their real-life counterparts. I further hypothesize that the characterizations of soap opera blacks are very similar to soap opera whites. Hence, their "blackness"—those unique qualities or speech styles that could identify them as black—are glossed over and are never really addressed.

Because of institutional racism, discrimination, and environmental and cultural differences, blacks and whites meet with different problems and situations throughout their lives. Blacks are different from whites. Black Americans display unique qualities attributable to their blackness.

> Looking at the Western Hemisphere as a whole, it is abundantly evident that many tangible elements of African ways, customs, attitudes, values and views of life survived the Atlantic crossing. In differing degrees, according to the complex social forces at work, numerous Negro communities in the Americas continue to draw from the African wellspring. [Courlander 1976:2]

Through extensive character and script analysis, this field project will evaluate whether blacks in soap operas exhibit any uniquely black characteristics or role situations. If unique situations are found, this will disprove the hypothesis that blacks are portrayed unrealistically on TV serials.

This topic was chosen because of the impact television can have on its viewers. Television gives the viewer more than mere entertainment; it relays images.

> The peculiar danger of television as a medium lies in the intensity of its impact. All television is subliminal advertising . . . the viewer continually registering in his mind situations which he may not realize he has seen. [Baggaley and Duck 1976]

These images are injected so subtly that they can be accepted without the viewer realizing that internalization has occurred. In this respect television can work as a detriment to society by promoting unrealistic images of racial or ethnic groups. Therefore, this field project will discuss some of the possible consequences of unrealistic presentations of blacks on TV in relation to contemporary American society.

This type of study is very important to anthropology because it examines the effects that a medium can have on American culture. Daytime television serials were selected because of the frequency with which they are shown and watched. More data can be extracted from a program that is shown five times a week than from those that are shown only once a week. A large viewing audience indicates that many viewers are at least exposed to the messages television serials transmit.

Operationalization of Hypothesis

It is important to define the critical terms that will be used in this project. A *daytime serial* (which will also be referred to as *soap opera* and *soap*) is a recurring television program that extends its characterizations and story lines (plots) over time, from one episode to another. This project will compare the situations of blacks on soaps to blacks in America by examining: (1) housing—does segregation exist on the soaps or in American culture? (2) unemployment—are the unemployment rates of blacks on soaps near the unemployment rates of blacks in America? (3) most importantly, occupational status and levels of occupational mobility—are these factors comparable for soap opera blacks and their real-life counterparts? An analysis of existing data on the present status of black Americans will be used to compare and contrast serial blacks and real-life blacks.

Low-status positions are defined as jobs where: (1) the position lacks any authority or management over other workers, *and* (2) the position requires no college education, technical skills, business skills, or apprenticeship before practicing in that profession; *or*, (3) the job requires committing an illegal act. Housewives were not treated as holding low-status positions.

The best way to collect and record data on soap opera characterization is by observing the programs on a regular basis. This project will draw on Susan Bean's (1976) delineation of three types of relationships: parentage, family relationships, and love triangles. It was necessary to limit the actual observation to programs broadcast on one network—the American Broadcasting Company (ABC)—for several reasons. First, since most television serials come on during the same hours, it would have been impossible to observe all the serials. Second, ABC is widely considered to have the most liberal story lines on television. Third, ABC has a larger black soap opera cast than any other network. Finally, I have watched the four ABC serials that will be used in this project—"General Hospital" (GH), "All My Children" (AMC), "One Life to Live" (OLTL), and "Ryan's Hope" (RH)—since they were first introduced to television. However, most of the older data are from memory and were collected before I observed these four shows for this anthropological study. Consequently, some of the data must be considered subjective. In an attempt to further support my subjective testimonies, four other avid viewers were asked to give a synopsis of the early plots that were relevant to this study.[1] Also, the plot synopsis section of the *Detroit Free Press* was used whenever issues were available to help follow present story lines. In order to help the nonwatcher through the maze of characters and situations that will be discussed, table 1 presents brief character descriptions of black cast members.

1. The four were asked open-ended questions related to sketching of various love triangles and biological kinship references from ABC television serials. The replies were all in agreement concerning the skeletal outlines of these relationships.

TABLE 1 Brief Character Descriptions of Black Cast Members

"Ryan's Hope" (RH)

Clem Lemphart (former black cast member)—Chief of neurosurgery at Riverside Hospital. No family or social life was ever presented in the story line.

"All My Children" (AMC)

Caroline Grant (former black cast member)—Nurse at Pine Valley Hospital. Former wife of Dr. Frank Grant.

Nancy Grant Blair—Administrative social worker. Widow of white social worker Carl Blair. They may have a *legal* son, Carl, Jr. His *biological* father is Dr. Frank Grant, Nancy's first husband.

Frank Grant—Doctor at Pine Valley Hospital. First wife Nancy, second wife Caroline. Illegitimate son conceived with Nancy while he was still married to Caroline but before Nancy had married Carl Blair.

Russ Anderson—Neurosurgeon at Pine Valley Hospital. Presently vying (along with Frank Grant) for the affections of Nancy Grant.

"One Life to Live" (OLTL)

Carla Gray Hall Scott—Executive secretary for the chief of staff at Llanview Hospital. Recently divorced Captain Ed Hall to marry Dr. Jack Scott.

Ed Hall—Captain of Llanview police force, presently running for Lt. Governor of the state. Carla Scott's first husband.

Josh Hall (former black cast member)—Juvenile delinquent who was adopted by Carla and Ed Hall. Sent away to school during his early teens.

Sadie Gray—Director of Housekeeping at Llanview Hospital. Mother of Carla Scott.

Jack Scott—Cardiac surgeon at Llanview Hospital. Recently gave up his playboy lifestyle to marry Carla Hall.

"General Hospital" (GH)

Claudia Larkin—Rich, upperclass college student who dates Brian Franklin.

Brian Franklin—Former teenage alcoholic, poor college student working on his Ph.D. in social work. Former roommate to Scotty Baldwin, a white attorney.

Presentation of Data and Results

Housing

Sociologist Barbara Carter used data supplied by the 1976 United States Bureau of Labor Statistics to describe the housing situation of black Americans.

> Especially significant is the fact that the races live almost entirely apart. . . . The extent of housing segregation has not changed since 1940. [Carter and Newman 1978:182]

Not only do all of the black soap opera characters reside in the same neighborhood as the white characters, whites and blacks often reside together. On AMC Caroline (black) and Donna (white) roomed together, as did Kristina (white) and Nancy (black). On GH Brian (black) and Scott (white) were roommates, and Ed (black) of OLTL stays at a boarding house with five whites. Although blacks do sometimes live with whites, it is unusual to see more than half of the black cast living with whites on television, while segregation runs rampant in contemporary American society. By ignoring segregation altogether, daytime television serials are somewhat unrealistic in terms of housing.

Unemployment and Occupational Status

Carter reports on the unemployment situation for black Americans.

> Unemployment rates are shockingly higher among Blacks than Whites and a larger proportion of Black people are living in poverty. Even comparable years of schooling and a college degree do not bridge the gap. [Carter and Newman 1978:183]

Tables 2 and 3 show an entire cast list for two ABC television serials.[2] Using the aforementioned definition of a low-status job, the tables show that 20 percent of OLTL's and 19 percent of AMC's white characters are either unemployed or have low-status jobs, whereas none of the blacks on either show is unemployed or holds a job with low status. This is extremely unrealistic because the unemployment rates for blacks are much higher than those for whites. Yet soap opera blacks not only have jobs, they have good jobs.

These statistics reveal two important concepts for this study. First,

2. "General Hospital" was not used because its two black characters are both college students. "Ryan's Hope" was not used because the serial no longer has a black character.

TABLE 2 Cast List for "One Life to Live"

Cast Member	Occupation	Managerial	Education
Katrina Karr	Former prostitute/ unemployed	no	no
Vicki Reilly	Heiress/former business person		
Karen Woleck	Former prostitute/ unemployed	no	no
Larry Woleck	Doctor		yes
Vincent Woleck	Police officer		yes
Ivan Kipling	Chief of Neurosurgery (M.D.)		yes
Faith Kipling	Housewife		
Mario Correlli	Doctor		yes
Edwina Lewis	Reporter		yes
Dorian Lord	Millionairess/former editor/ former M.D.		yes
Pat Ashley	Talk show hostess		yes
Maggie Ashley	Executive secretary		yes
Peter Jaansen	Doctor		yes
Bradley Vernon	Entrepreneur	yes	
Jenny Vernon	Nurse		yes
Samantha Vernon	Entrepreneur	yes	
Will Vernon	Chief of Staff (M.D.)	yes	yes
James Craig	Retired (M.D.)		yes
Anna Craig	Housewife		
Fran Gordon	Unemployed		
Mick Gordon	Handyman	no	no
Herb Callison	District Attorney	yes	yes
Clint Buchannon	Newspaper editor	yes	
Becky Abbot	Aspiring singer	no	no
Ina Hopkins	Landlord	yes	
*Carla Hall Scott	Executive secretary		yes
*Jack Scott	Cardiac surgeon		yes
*Ed Hall	Police captain/running for Lt. Governor	yes	
*Sadie Gray	Housekeeping director	yes	

*Black cast member

TABLE 3 Cast List for "All My Children"

Cast Member	Occupation	Managerial	Education
Joe Martin	Doctor		yes
Ruth Martin	Nurse		yes
Kate Martin	Matriarch/housewife		
Paul Martin	Attorney		yes
Anne Martin	Currently insane		
Phoebe Tyler	Millionairess		
Charles Tyler	Chief of Staff (M.D.)	yes	yes
Chuck Tyler	Doctor		yes
Donna Tyler	Former prostitute/ housewife		
Lincoln Tyler	Attorney		yes
Kelly Cole	Addict/singer/unemployed		
Myrtle Fargate	Salesperson	no	no
Claudette Montgomery	Hostess-Asst. Manager		
Langley Wallingford	Con artist		
Eddie Durantz	Deceased con-artist		
Edna Sego	Manicurist	no	no
Benny Sego	Chauffeur	no	no
Maristella Tuggle	Former prostitute/ unemployed		
Billy Clyde Tuggle	Pimp/drug pusher		
Mark Dalton	Asst. Professor		yes
Erica Cudahey	Entrepreneur	yes	
Mona Kane	Executive secretary		yes
Tom Cudahey	Entrepreneur	yes	
Bruno	Bartender	no	no
Ellen Shepherd	Entrepreneur	yes	
Betsy Kennicott	Student nurse		yes
Sybil Thorne	Student nurse		yes
Cliff Warner	Doctor/intern		yes
Palmer Courtlandt	Entrepreneur	yes	
Nina Courtlandt	Student		yes
Brooke English	Student/heiress		yes
Harlan Ricker	Retired/businessman	yes	
*Frank Grant	Doctor		yes
*Nancy Grant Blair	Administ. social worker	yes	yes
*Caroline Grant	Nurse		yes
*Russ Anderson	Neurosurgeon		yes

*Black cast member

there is a definite, obvious lower class on daytime television serials. Not all of the characters are professionals. Soaps might partially justify the high occupational status of television blacks were everyone in the cast a professional. But in the last five years ABC soap operas have incorporated a lower class into their basic schemes. Second, this appearance of the lower class makes the high occupational status of blacks even more striking. For example, the four black characters on "All My Children" together have twenty-six years of college! "Ryan's Hope" no longer has a black cast member, but the only black character the show ever had was the chief of neurosurgery at Riverside Hospital. The majority of blacks in America do not hold the occupational positions that their serial counterparts do.

The characterizations of blacks in relation to their professional capabilities is unbelievable. Several of these characters show great mobility. Jack Scott, Frank Grant, and Price Trainer (a former black cast member) all worked their way out of the ghetto to become prominent doctors. Dr. Scott is supposed to be one of the best cardiac surgeons in the country. Carla Scott's mother, Sadie Gray, worked her way up from being a housemaid to the head of hospital housekeeping within three seasons. Brian Franklin was a poor teenage alcoholic who worked his way out of poverty and into graduate school. Also, the viewer is seldom allowed to witness this mobility because in 80 percent of the present cases, blacks entered the story line *after* they had achieved such high levels of upward mobility. The level of mobility on the soap operas is thus very unrealistic in comparison to contemporary American society.

> Except for a small number of success stories, Blacks remained America's underclass, mired in poverty. . . . 200 years after the Declaration of Independence, Blacks were still unable to benefit from upward mobility. [Ploski and Marr 1976:468]

In terms of role expansion, the size of black roles has increased and the number of spoken words has increased. However, the characters' "blackness"—those qualities or speech styles that identify the character as black—has not increased. For example, in 1969 when Carla Scott was first introduced as a character, her blackness had to be dealt with because she was pretending to be white. After it was discovered that she was black, discussions about race almost totally disappeared from the script. Carla's gradual acceptance of her blackness could have allowed the character to examine all areas of black life for the first time on TV because she had been "passing" all of her life. Instead, Carla smoothly moved from white to black life without an analysis of the difference between the two.

One uniquely black characteristic that is never approached on ABC serials is dialect. The characters who come from the ghetto don't exhibit patterns of speech that are representative of ghetto life. None use a type of dialect known as "Black English," which is prevalent among contemporary ghetto blacks. I am not asserting that all ghetto blacks use Black English, but it is an accepted fact that most do use some form of Black English, even if they have left the ghetto. Certainly if, regardless of their backgrounds, all of the black characters spoke Black English, this would be stereotypical. However, stereotypes do have some degree of truth. It is equally unrealistic to present *all* soap opera blacks as using standard American English, because some blacks do use Black English.

Some of the black roles have not been expanded to the extent that the black character has a personal and family life. For example, the four ABC serials under study presently have only one black child and one married black couple. Sadie Gray has not had a date, lover, or spouse since her role was introduced years ago. Generally, black roles have not reached the level of complexity that many white characters' personal lives have. Instead, blacks are often shown in the capacity of friends or co-workers. For instance, Brian Franklin and Claudia Larkin usually share their GH scenes with their white best friends, Laura and Scotty Baldwin. Since I have been observing this show throughout November, Brian and Claudia have had three scenes alone. In the three scenes the bulk of the discussion centered around the Baldwins' marital problems. Claudia and Brian have been dating for over seven months. Throughout my observations, I have never seen this couple kiss. In the world of soap operas, whirlwind romances happen frequently, and they often occur consecutively. Within this same time span of seven months, white doctors Rick Webber and Monica Quartermaine separated from their spouses, had an affair, and reunited with their spouses. Mitch Williams had an affair with Tracy Quartermaine, then had an affair with Susan Moore, later married Tracy Quartermaine, and is presently having another love affair. The point is obvious. The personal lives of black characters move very slowly in comparison to those of their white soap opera counterparts.

The few black roles that have been expanded follow the same plot lines as white characterizations. On AMC, Frank and Nancy Grant were married. They were later divorced. While Frank was married to Caroline he conceived an illegitimate child with Nancy. Nancy then married Carl Blair on his deathbed and promised never to reveal the child's true parentage. She raised the child as Carl Blair, Jr. Frank wanted to tell Carl, Jr., that he was the boy's real father while Nancy demanded that Frank keep quiet. This is a very familiar situation to AMC viewers. A few years earlier, white characters Chuck, Phil, and Tara had a similar problem. Philip and Tara conceived an illegitimate child. Tara married Chuck and they raised the child as Chuck's son. Philip wanted to tell his son of his proper parentage while Tara vehemently protested that the truth should be kept a secret.

One of the most overused story lines in daytime television serials is the love triangle. After four years of happy marriage, the lone black couple on OLTL was placed in a love triangle. Carla and Ed Hall were married. Carla fell in love with Jack Scott and divorced Ed to marry Jack. This same type of love triangle has ruined many white marriages on ABC soap operas. On GH, Jeff and Monica Webber were married. Jeff divorced Monica to marry Heather. Lamont and Katie Corbin were married. Katie fell in love with Marc Dante and divorced Lamont to marry Marc. On AMC, Phil and Erica Brent were divorced so that Phil could marry his true love, Tara. These characterizations show that there is no difference between black and white soap opera characterizations. They have similar housing and jobs and when black roles are finally expanded, they are merely repeats of white personal relationships.

There is very little innovation used in terms of addressing the "blackness" of black soap opera characters. The only area where race is emphasized is that of interracial romantic relationships. Four interracial relationships have occurred on daytime serials. Three were very brief, while the fourth developed over two years on an National Broadcasting Corporation (NBC) show called "Days of Our Lives" (DOOL). In the first relationship, a black female doctor and a blind white patient began to fall in love. Shortly thereafter, both characters left the show and the relationship ended. Three years ago, Carla and Ed Hall adopted a black adolescent who eventually began "seeing" a young white girl. The relationship was never resolved because the black character was written out of the story line by sending him away to school. Last year Nancy Grant married Carl Blair on his deathbed after an airplane crash. Nancy was pregnant by her first husband, a black doctor, at the time of Carl's death. She had never slept with Carl Blair.

The most famous interracial love relationship led to engagement on

"Days of Our Lives." David Banning (white) and Valerie Grant (black) fell in love over a two-year span and were to be married, although the couple had only been allowed to kiss once on the show. The couple parted when Val was given an out-of-state scholarship to attend medical school.

The data on interracial relationships show several things. First, while physical contact often occurs among couples of the same race, it is kept at a minimum among interracial couples. This is unrealistic because the viewing audience has grown to expect characters who are developing relationships to kiss and to have some type of physical contact. Second, interracial couples do not stay together. The relationships are ended before the characters are able to address the sociological ramifications of interracial relationships. This is not representative of contemporary American society, where interracial marriages have increased over the past twenty years. Finally, in three of the four cases the black character was written out of the serial. This is a serious problem because there are so few black soap opera characters to begin with. Furthermore, the dismissal of black characters can be implicitly understood (or misunderstood) by some viewers to be a punishment for attempting to delve into an interracial romance. Television can intentionally or unintentionally transmit messages of this sort to viewers.

The most thought-provoking scene I observed was between black doctor Frank Grant and his white friend Chuck. Frank and Chuck were discussing the similarities of parentage problems between Frank's son and Chuck's son. Frank explained to Chuck that the situations were different because Carl, Jr., will grow up believing his father was white and his mother is black whereas both of his parents are black. Frank said this was important because he wanted his child to grow up knowing he was born of black parents. The crucial point is that Frank did not explain *why* it was important that the child know he is black. What are the social differences of a black child growing up thinking he is the product of a mixed marriage? This situation gave Frank Grant a chance to assert his blackness, to explain what it means to be black and why a child would need to know that he was black. In the scene Chuck went on to argue that the child's race was not an issue and that Frank was merely using race as an excuse to tell the child his true parentage. The scene mentioned race but it did not address or confront the issue.

The consequences of the unrealistic portrayal of blacks on soap operas cannot be fully assessed until the impact of television is understood.

> Once considered a passing fancy, television has become an important agent of socialization for both children and adults, conveying a vast array of attitudes and behavior. Ironically television may be at its most powerful when it is presenting "mere" entertainment. . . . Important lessons are transmitted, often incidentally, and subtly. [Donogher et al. 1975:1031]

If this assertion seems farfetched, consider the fact that many of those who view soap operas are somewhat isolated in their homes during the afternoons. These viewers are allowed to intrude on the most intimate aspects of personal relationships. The more the viewer becomes entranced with the situation, the more real and believable the characters become. These characterizations teach isolated viewers about groups from which they are isolated. One such group is the black community.

What are the lessons TV soap operas transmit about blacks? From this study, it can be concluded that television serials transmit unrealistic images of blacks with inflated occupational mobility and increased employment. These serials imply that interracial relationships do not last and that all blacks speak standard English.

What are some of the possible consequences of this unrealistic portrayal? The most obvious is that soap opera viewers will have misconceptions about the situation of blacks in America. Viewers who accept soap opera images would not understand any collective actions designed to uplift the positions of blacks because serials transmit the message that conditions for blacks are good and stable. Furthermore, viewers who have incorporated the soap opera images of the black man and woman being able to work their way out of poverty and into the operating rooms would not understand the necessity of some mobilizing and compensatory programs, such as affirmative action. The final possible effect could be the loss of support for worthwhile and necessary programs designed to help contemporary blacks reach the plateau that their soap opera counterparts already occupy.

Human Behavior (1974:46) discussed the effect of such unrealistic black characterizations on the black community.

> Black residents of Hearthrob city are seldom around for long; most "come and go." None are either unemployed or on welfare. One can applaud the effort to show the white audience that blacks can achieve like whites. But does it promote racial pride in black viewers or frustration and anger?

It would be very difficult for blacks who have internalized the black soap opera situation as reality to justify or blame their unemployment or lack of mobility on anyone other than themselves. For if individuals like Brian Franklin, a teenaged, poverty-stricken alcoholic, can pull himself out of the ghetto and into college, surely any black American can do so. Furthermore, by showing so many blacks with professional degrees, soap operas imply that success is related to a professional degree. Hence, blacks cannot feel secure and proud because they have achieved some level of mobility, such as becoming a mechanic, because black success is indicated by a professional degree. *Human Behavior* (1974:46) substantiates this point.

> Soaps help to promote unrealizable expectations—among them, the notion that a professional degree is the only assurance of self-respect and happiness . . . ignored skilled trades and service occupations.

Summary

This content analysis has shown that black characterizations are similar to those of whites, while the character analysis showed that soap opera blacks are in a much better position than their real-life counterparts. If ABC serials are the most innovative on daytime television, then these conclusions should be supported by observational studies of other networks.

This type of study could be handled better by a group of observers taking notes or taping serials on a daily basis over a longer time. Using only one major observer was not conducive to good data retrieval because I was not always able to watch three and one-half hours of television per day. Also, persons of various ages, races, and sexes should be used to help locate any bias that might distort the results. Although my observations took an etic perspective, it would be impossible for me to say that my race did not affect my perceptions of situations viewed prior to observation specifically for the study.

I see the major problem with the television serials to be defining "none," "some," and "all." Surely there are *some* blacks who do not have low-status occupations, but it is unrealistic to show *no* blacks with low-status positions. The same is true about Black English; it appears that in an attempt to compensate for past stereotypical roles, television serials have overcompensated by conspicuously removing *all* blacks from the lower class. This is an example of Lévi-Strauss's binary oppositions; blacks have either low status or high status. Either she is a maid, Sadie Gray's first job, or the head of housekeeping, Sadie's present position. Jack Scott began in poverty. He is now one of the best cardiac surgeons in the country. The poverty from which the characters came is sometimes mentioned, while the wealth and status which they have achieved is seen through conspicuous consumption. A mediating position is presently missing from black life on television serials.

In this final student paper, Fermin Diez, who is Venezuelan, comments on Americans' fascination with sports. He identifies some of the old and new American values expressed in our spectator sports and links the increasing relative popularity of football over baseball to recent changes in American society. This selection by Diez brings a needed foreigner's perspective to the interpretation of American culture. Diez draws on the mass media, conversation with natives, and his own observation and experiences while living in the United States in his article, which is more journalistic than most of the other papers presented here.

25 / **The Popularity of Sports in America: An Analysis of the Values of Sports and the Role of the Media**

Fermin Diez

One of the most striking things about the mass media in America is the extensive coverage that sports events receive. Most major newspapers include a section devoted to sports (and many Americans rarely read more than this; cf. Cozens and Stumpf 1953). Both radio and television news dedicate 25 to 33 percent of their time to sports. However, extensive sports coverage can best be appreciated in television programming. More than 50 percent of national prime-time broadcasting (6:00 P.M. through 11:30 P.M. weekdays, and noon to 11:30 P.M. weekends) between September 16, and December 1, 1979, was given to such events as baseball play-offs, the World Series, college and professional football, professional basketball, tennis, bowling, boxing, and golfing. (The data were compiled from *TV Guide*.) This is an impressive statistic when we consider that only about 20 percent of prime-time programming consisted of news and cultural programs. (Remember, too, that news broadcasts also include sports.) It is estimated that "79 percent of all the households in the country tuned in the first Superbowl on TV" (Arens 1976). And, as has been pointed out, nearly always during the commercial breaks of major sports telecasts the water pressure all over America drops as toilets are flushed or water is drunk. Why do so many Americans watch so much sports programming? Why do Americans watch four football games on fall weekends, or nine games in five days during the Thanksgiving holidays? How do sports captivate such a vast and diverse audience? How are teams perceived by fans? And why do some players get such fame, glory, wealth, and social status? What American values are expressed in sports? These are some of the questions that will be examined in this paper. My data come from personal observations, from participant-observation, and from various sources in the anthropological and sports literature.

Conrad Kottak (1978*a*:508) contends that an interest in sports

> unites Americans regardless of ethnic group; region; state; urban, suburban or rural residence; religion; political party; job; status; wealth; gender; or sexual preference.

Sports attract an audience ranging from former presidents (e.g., Gerald Ford, who played football as an undergraduate) to individuals in the lower socioeconomic classes. On the other hand, males are, by and large, far more interested in sports than are females. This is related to sex-role stereotyping characteristic of network programming. TV shows aimed at women consist mostly of soap operas, which are aired in early afternoon, a time when women are presumably at home and men at work. Sports, however, are aired mainly during prime time, when it is assumed that both men and women are at home. It is my belief that the networks presume that men have priority in choosing shows, and that they choose sports; this reinforces male dominance. Furthermore, networks not showing sports events during time periods when others are show programs geared more toward women. (For example, "Women to Women" was broadcast by ABC on Sundays at 5:30 P.M., while NBC and CBS were showing professional football.) Since male attention is already taken, potential viewers are mainly women, and in this case more independent women.

Although it is assumed that television networks are "in it for the money," William C. MacPhail, former vice president of CBS-TV Sports, contends

> We do great if we break even. Sports is a bad investment, generally speaking. The network needs it for prestige, for image, to satisfy the demands, the desires of our affiliated stations. The rights have gotten so costly that we do sports as a public service rather than a profit maker.

The fact that NBC paid nearly $250 million for the rights to the 1984 Summer Olympics in Los Angeles makes MacPhail's story more credible. There must clearly be something here that hits a major chord in American culture; and the most obvious place to look for this something is in the values contained in and portrayed by big time sports, particularly baseball and football. Let us examine these two sports in order to see what can be discovered about American values.

Baseball

Baseball is a relaxed sport, in which action occurs sporadically; this permits thinking in between. Since plays and rules are uncomplicated, most people can follow events and can call or predict plays. In fact, baseball plays with a proverbial "book," or standard of playing "good baseball"; when things are not done according to this book, managers or players are thought "crazy" or "unorthodox." Baseball thus allows spectators to feel on top of things, to know and to follow a game without having to pay much attention. Unlike football or basketball, baseball could probably be defined as a team sport played by individuals. Each player depends mostly on himself, on his own ability to catch, throw, or hit the ball. "An emphasis on individual responsibility" (Arensberg and Niehoff 1975) is a well-known trait, and it shows up in baseball.

Baseball depends as much on luck as on skill. Thus a player may hit the ball well, but right into the center fielder's glove. A badly hit ball may drop between the shortstop and the left fielder for a "Texas leaguer." A batter may make contact with an excellent breaking pitch by chance and hit the game-winning home run, or a bad hop may elude a skilled second baseman. Consequently, baseball includes significant rituals and "magic" (Gmelch 1975). Players will (magically) do things in exactly the same way if they think that it will help them perform well. As Gmelch (1975:349) points out,

> Detroit Tiger infielder Tim Maring wore the same clothes and put them on exactly in the same order each during a batting streak.

Conversely, players believe that if they don't do these things, they will not play well. Both luck and skill determine the outcome of a game. This is probably also why a seven-game series is used to determine the best baseball team of the season as opposed to the all-or-nothing Super Bowl in football.

What, however, do traditional values have to do with the popularity of baseball in today's highly technological America? That is precisely the point—nothing! Fifty years ago, when such immortals as Babe Ruth played the game, these values were important, since people felt (particularly during the Depression) that their lives were not really in their own hands, and that luck played a large role in human affairs. Baseball players defied luck in every play, and succeeded; the long-time reign of the New York Yankees testifies to that. But as technology came to rule people's lives more and more, baseball's popularity declined, so that it is no longer the "national pastime," leaving the door open for football.

Football

Football is definitely a team sport; each player must do his assignment well for the plays to be executed properly. No pass or run would succeed without blockers. Specialization is a key feature of football. We have quarterbacks, halfbacks, tailbacks, receivers, place kickers, punters. There are even specialty units for kickoffs, field goals, and punt returns. Furthermore, luck's role is reduced; if every job is done efficiently, the play is completed. The other component of the nature side of the nature/culture opposition in baseball (the field) is less important in football than is the clock, a cultural component. It does not require much imagination to see parallels between football and today's American way of life, with its dominant business corporations and industrial organizations, in which each individual has a specialized job that must be completed in order for the system to go on efficiently, and in which technology has set the standards of specialization.

Competition is another major part of this American way of life, and is represented in football by the offense/defense opposition. If you do not do your job well, the other team will take advantage of it, and the trick is to do your job better than the opposing team. You must nullify their offense and overpower their defense; otherwise they will do it to you. Is this not what competition among industries and corporations is all about? Football's complexity thus resembles in many ways the complex life of Americans today, and football games reassure spectators that business/industry really works (Montague and Morais 1976).

Many other American cultural traits also appear in football. We see individualism in praise for the running back who rushes for more than 100 yards, for the quarterback who passes for a touchdown, for the kicker who makes a 45-yard field goal, for the lineman who sacks the quarterback five times. Leadership is another value, personified by the quarterback, the most prized player on the team. Hierarchy shows up in football, as decisions come from above (the coach), through executives (the quarterback and the defensive team captain), all the way down to the blue-collar workers (the linemen). This again is reminiscent of corporations. And there is the American value of aggression. Thus, baseball is an American tradition that persists, like hot dogs and apple pie, whereas football is the idealization of the actual American way of life as portrayed in sixty minutes by two teams.

This leads us to the next question: Why do so many Americans watch football and baseball, and prefer the former over the latter? Part of the answer has been given above. These sports resemble the American way of life and portray some of its values. Still other values are involved: achievement, physical fitness, hard work. Rooting for a team provides social identity. Let us examine some of these other values in order to better understand the spectator side of the game/spectator dichotomy, so we can later draw the general conclusions and answer the questions raised in the introduction.

American Values

Achieving is a major American value, as Arensberg and Niehoff (1975) point out, and athletes are definitely achievers. Most schoolboys play sports and dream of becoming stars; professional athletes personify that dream, and are thus idolized by fans who admire physical prowess, another value of Americans (cf. Miner 1975). But in order to reach such a standard, many hours have to be spent running, lifting weights, practicing; natural talent alone is usually not sufficient for a 235-pound body and the ability to sprint sixty yards in less than five seconds.

Hard work is another American value, as the oft-quoted "practice makes perfect" points out. Furthermore, athletes receive fame, prestige, social status, and wealth. Some examples: Pete Rose, Joe Namath, Willie Stargell, and Nolan Ryan.

Probably more important than the way individual athletes are seen by American fans is the way in which teams are perceived, and the identification of fans with this team image. The New York Yankees, for example, have a "WASP" image. The Dallas Cowboys are efficient and businesslike; the Pittsburgh Steelers are the "rough" and "macho" team in football. The Oakland A's, during their three world-championship seasons, were the "swingin' A's." The Cincinnati Reds are middle American family men. Notre Dame owes its large national following to its Catholic image. Teams tend to attract fans who see themselves in terms of the image portrayed by a particular team, and thus identify with it.

This introduces the next value—the importance of winning versus the despair of losing. The old saying, "It's not whether you win or lose, but how you play the game," was never intended for big time sports, in which, as Vince Lombardi once put it, "winning isn't everything, it's the only thing." Few remember who came in second, whereas any avid sports fan can name the teams that won the pennant for ten or twenty seasons. Another example of the importance of winning is criticism of the University of Michigan's head coach "Bo" Schembechler for losing bowl games, despite the fact that his teams have not had a losing year and have won numerous Big Ten championships. Similarly, "Sparky" Anderson was fired by the Cincinnati Reds because, after five National League pennants and two world championships, he failed to finish the next season at the top of the division. The 1979 Baltimore Orioles seemed a bit of a fraud when, after winning over a hundred games, they lost the World Series.

Losing brings a devastating sense of despair to the fans. The depression of University of Michigan students after their 1979 home loss to arch rival Ohio State provides a good example. But this is nothing compared to the Denver Broncos fan who shot himself in the head after his team lost to the Chicago Bears by fumbling seven times. There is, however, one fact that seems not to fit this analysis—the "underdog" syndrome, rooting for a team unlikely to win. The explanation for this is actually quite simple: not everybody is a winner. Since sometimes we are all losers, it is only logical that we sometimes identify with losing teams.

The media, as the name suggests, act as mediators between the game and the spectators. Every World Series, Super Bowl, weekend, holiday, night, or whatever, the spectator sits comfortably in front of the TV watching the fate of his or another team. TV has created a "sports hype" through the exorbitant sums it pays for the rights to transmit the games. This has financed the expansion of many teams to western cities (Los Angeles, Oakland, etc.). Televised sports now unite the whole country with sports in all regions. All of the American values discussed above are accentuated by the media. TV stresses these values not just by the time allotted to sports, but by commenting on or drawing attention to individual or team performance. The best example of this is Howard Cosell, who appears to feel no remorse at cutting down a poor performance, as he did in a 1979 Monday night game in which the Seattle Seahawks routed the New York Jets. The Jets had played "a sad excuse of a professional football game." Losers and losing are bad, and Cosell communicates this message to millions of spectators all over the country. Yet he also seems never to tire of praising a good performance, e.g., by

Walter Payton, "the best NFL rusher . . . and possibly one of the best who ever played the game." Payton led the Chicago Bears to victory over the Minnesota Vikings, with nearly two hundred yards rushing. Again the message is clear: winners are good, losers are not.

But there are still other ways in which TV stresses these values. Anyone who has watched a sports event on television has noticed the individualized comments, usually statistics, that most players get as the camera does a close-up. Half-time highlights show outstanding plays, or ridicule blunders. As the season ends, TV networks show only teams who are on top of their divisions. All of this serves to reinforce the image of players and teams as winners or losers. During the preseason, spectators watch workouts and training camps stressing hard work. Other statistics, such as size, weight, and speed of a player, stress physical fitness. TV also aids in creating team images with statements like: "the colorful Oakland A's," or "the grinding Pittsburgh Steelers," or "the intelligent coaching of the Cowboys." Also, prestige is acquired by players like Tom Seaver who act as commentators in important sports telecasts. TV turns athletes into heroes, e.g., James Scott, a light-heavyweight boxer and a convict in a New Jersey prison who was allowed (thanks to TV dollars!) to fight professionally in the prison's courtyard; his fights (mostly arranged and scheduled by ABC) wound up on national television.

From this analysis of American values present in big time sports, ways in which spectators identify with these values, and the impressive coverage sports events receive nationwide, we see some of the reasons why sports have become such a powerful element in the daily lives of Americans. Sports feed a need in Americans to know that their values have results and that both tradition and technology are good. The media, especially TV, simply satisfy this demand.

References Cited

Ann Arbor Bank and Trust
1979 Welcome to AAB. *In* Employee Handbook.

Anonymous
1974 The Good Life of the Soaps. Human Behavior 3 (January):62–63.

Anonymous
1979 Hilarious Haiku. Detroit Magazine, November 25, p. 5.

Arens, William
1976 Professional Football: An American Symbol and Ritual. *In* The American Dimension: Cultural Myths and Social Realities. William Arens and Susan Montague, eds. pp. 3–14. Port Washington, N.Y.: Alfred Publishing Co.

Arens, William, and Susan Montague, eds.
1976 The American Dimension: Cultural Myths and Social Realities. Port Washington, N.Y.: Alfred Publishing Co.

Arensberg, Conrad, and Arthur Niehoff
1975 American Cultural Values. *In* The Nacirema. James P. Spradley and Michael A. Rynkiewich, eds. pp. 363–378. Boston: Little, Brown, and Co.

Assagioli, Roberto
1973 The Act of Will. New York: Viking Press.

Baggaley, Jon, and Steve Duck
1976 The Dynamics of Television. Farnborough, Hants.: Saxon House.

Bateson, Gregory
1958 Naven. Stanford: Stanford University Press.
1972 Steps Toward an Ecology of Mind. New York: Ballantine Books.

Bather, Francis Arthur
1977 The Puns of Shakespeare. *In* Crosbie's Dictionary of Puns. John Crosbie, ed. pp. 267–278. New York: Harmony Books.

Bean, Susan
1976 Soap Operas: Sagas of American Kinship. *In* The American Dimension: Cultural Myths and Social Realities. W. Arens and S. Montague, eds. pp. 80–98. Port Washington, N.Y.: Alfred Publishing Co.

Benet, William
1965 The Reader's Encyclopaedia. 2nd ed. New York: Thomas Crowell Co.

Besmajian, Haig
1974 The Language of Oppression. Washington, D.C.: Public Affairs Press.

Bettelheim, Bruno
1975 The Uses of Enchantment: The Meaning and the Importance of Fairy Tales. New York: Alfred A. Knopf.

Boston Women's Health Book Collective
1976 Our Bodies, Ourselves. New York: Simon and Schuster.

Brownmiller, Susan
1976 Against Our Will: Men, Women and Rape. New York: Bantam Books.
Burchinal, Lee G.
1963 Personality Characteristics of Children. *In* The Employed Mother in America. F. Ivan Nye and Lois Wladis Hoffman, eds. Chicago: Rand McNally.
Carleton, G. W., and Co.
1877 Carleton's Handbook of Popular Quotations. New York: G. W. Carleton and Co.
Carter, Barbara, and Dorothy K. Newman
1978 Perceptions About Black Americans. The Annals of the American Academy of Political and Social Science 435:179–205.
Chesler, Phyllis
1972 Patient and Patriarch: Women in the Psychotherapeutic Relationship. *In* Woman in Sexist Society. Vivian Gornick and Barbara K. Moran, eds. pp. 262–292. New York: New American Library.
Ciriacy, Edward, and Lowell Hughes
1973 Contraceptive Counseling. *In* Family Practice. Howard F. Conn, ed. pp. 298–309. Philadelphia: Saunders.
Corea, Gena
1977 The Hidden Malpractice. New York: William Morrow.
Courlander, Harold
1976 A Treasury of Afro-American Folklore. New York: Crown Publishing Co.
Cozens, Frederick W., and Florence Scovill Stumpf
1953 Sports in American Life. Chicago: University of Chicago Press.
Crosbie, John, ed.
1977 Crosbie's Dictionary of Puns. New York: Harmony Books.
Daley, Robert
1977 The Search for Sam: Why It Took So Long. New York Magazine, August 22, 1977.
Deutsch, Helene
1944 The Psychology of Women. Vol. 1. New York: Stratton.
Dixon, Franklin W.
1935 The Hidden Harbor Mystery. New York: Grosset and Dunlap Co.
1961 The Hidden Harbor Mystery. 2nd ed. New York: Grosset and Dunlap Co.
Donogher, P., et al.
1975 Race, Sex and Social Example. An Analysis of Character Portrayals in Interracial Television Entertainment. Psychological Reports 37:1023–1034.
Durkheim, Emile
1954 The Elementary Forms of the Religious Life. J. W. Swain, transl. New York: Free Press (1st ed. 1915).
Ember, Carol, and Melvin Ember
1973 Cultural Anthropology. Englewood Cliffs, N.J.: Prentice-Hall.

Emerson, Joan
1973 Negotiating the Serious Import of Humor. *In* People in Places. A. Birenbaum and E. Sagarin, eds. pp. 269–280. New York: Praeger Publishers.
Employer's Insurance of Wausau
N.d. The Feminine Touch. N.p.
Ervin-Tripp, Susan
1973 An Analysis of the Interaction of Language, Topic, and Listener. *In* Language Acquisition and Communicative Choice, pp. 239–259. Stanford: Stanford University Press.
Farb, Peter
1977 The Unspeakable Pun. *In* Crosbie's Dictionary of Puns. John Crosbie, ed. pp. 281–282. New York: Harmony Books.
Fikes, Jay Courtney
1978 Native American Education: Cognitive Styles, Cultural Conflict, and Contract Schools. Michigan Discussions in Anthropology 4(1):31–51.
First Federal Savings & Loan Association of Battle Creek
1979 Employee Manual.
First Federal Savings of Detroit
1979 The First Federal Story. N.p.
Fiske, Shirley
1975 Pigskin Review: An All American Initiation. *In* The Nacirema. James P. Spradley and Michael A. Rynkiewich, eds. pp. 55–68. Boston: Little, Brown, and Co.
Gal, Susan
1978 Peasant Men Can't Get Wives: Language Change and Sex Roles in a Bilingual Community. Language in Society 7:1–16.
Gard, Larry
1978 Banks—A Ritual Institution. Manuscript, University of Michigan Department of Anthropology, Ann Arbor.
Gingold, Judith
1976 One of These Days—Pow—Right in the Kisser. Ms. Magazine 5 (February):51–54, 94.
Gmelch, George J.
1975 Baseball Magic. *In* The Nacirema. James P. Spradley and Michael A. Rynkiewich, eds. pp. 348–353. Boston: Little, Brown, and Co.
Goldberg, Philip
1968 Are Women Prejudiced Against Women? Trans-action 5:28–30.
Gordon, Harvey
1980 PUNishment. 2nd ed. New York: Warner Books.
Graebner, Alan
1975 Growing Up Female. *In* The Nacirema. James P. Spradley and Michael A. Rynkiewich, eds. pp. 23–29. Boston: Little, Brown, and Co.
Graham, Alma
1975 The Making of a Nonsexist Dictionary. *In* Language and Sex. Barrie Thorne and Nancy Henley, eds. pp. 57–63. Rowley, Mass.: Newbury House.

Greenhill, Jacob Pearl
1965 Office Gynecology. 8th rev. ed. Chicago: Yearbook Medical Publications.

Gross, Daniel R.
1971 Ritual and Conformity: A Religious Pilgrimage to Northeastern Brazil. Ethnology 10:129–148.

Gulliver, P. H.
1974 The Jie of Uganda. *In* Man in Adaptation: The Cultural Present. 2nd ed. Y. A. Cohen, ed. pp. 323–345. Chicago: Aldine.

Harris, Marvin
1974 Cows, Pigs, Wars and Witches. New York: Random House.
1979 Cultural Materialism: The Struggle for a Science of Culture. New York: Random House.

Henry, Jules
1975 Golden Rule Days: American School Rooms. *In* The Nacirema. James P. Spradley and Michael A. Rynkiewich, eds. pp. 30–43. Boston: Little, Brown, and Co.

Herschberger, Ruth
1970 Adam's Rib. New York: Harper and Row.

Hewitt, John
1976 Self and Society. Boston: Allyn and Bacon.

Hirschmann, Lynette
1974 Analysis of Supportive and Assertive Behavior in Conversations. Paper presented to the Linguistic Society of America.

Hollingshead, August de Belmont
1949 Elmtown's Youth. New York: John Wiley.

Holmes, Oliver Wendell
1960 The Autocrat of the Breakfast-Table. London: J.M. Dent and Sons. (1st ed. 1858).

Hsu, Francis L. K.
1975 American Core Values and National Character. *In* The Nacirema. James P. Spradley and Michael A. Rynkiewich, eds. pp. 23–30. Boston: Little, Brown, and Co.

Huntington, Richard, and Peter Metcalf
1979 Celebrations of Death. New York: Cambridge University Press.

Janeway, Eliot
1977 Reviving the Economy: If Women Can't Do It No One Can. Working Woman 2(10):66–67.

Kaplan, David, and Robert A. Manners
1972 Culture Theory. Englewood Cliffs, N.J.: Prentice-Hall.

Kardiner, Abram
1954 Sex and Morality. Indianapolis: Bobbs-Merrill.

Kottak, Conrad P.
1978*a* Anthropology: The Exploration of Human Diversity. 2nd ed. New York: Random House.
1978*b* Social Science Fiction. Psychology Today 11(9):12, 17, 18, 106.

1980 The Past in the Present: History, Ecology, and Cultural Variation in Highland Madagascar. Ann Arbor: University of Michigan Press.

Lakoff, Robin

1973 Language and Woman's Place. Language in Society 2:51–53.

Lamb, Charles

1977 The Last Essays of Elia, Popular Fallacies—IX. *In* Crosbie's Dictionary of Puns. John Crosbie, ed. pp. 279–280. New York: Harmony Books.

Larkin, Ralph W.

1979 Suburban Youth in Cultural Crisis. New York: Oxford University Press.

Lee, Dorothy

1950 Codification of Reality: Lineal and Nonlineal. Psychosomatic Medicine 12(2):89–97.

Lenanne, K. J., and R. Lenanne

1973 Alleged Psychogenic Disorders in Women—A Possible Manifestation of Sexual Prejudice. New England Journal of Medicine 288(6):288–292.

Lévi-Strauss, Claude

1966 The Savage Mind. Chicago: University of Chicago Press.

1967 Structural Anthropology. C. Jacobson and B. G. Schoepf, transl. Garden City, N.Y.: Doubleday.

1969 The Raw and the Cooked. New York: Harper and Row.

Lundberg, Ferdinand, and Marynia F. Farnham

1947 Modern Woman: The Lost Sex. New York: Harper and Row.

McHugh, Peter

1968 Defining the Situation. Indianapolis: Bobbs-Merrill.

MacNeish, Richard

1967 Introduction. *In* Prehistory of the Tehuacan Valley, Vol. 1. D.S. Byers, ed. pp. 3–13. Austin: University of Texas Press.

Malinowski, Bronislaw

1961 Argonauts of the Western Pacific. New York: Dutton.

Margolis, Maxine

1977 From Betsy Ross Through Rosie the Riveter: Changing Attitudes Towards Women in the Labor Force. Michigan Discussions in Anthropology 3(1):1–40.

Miner, Horace

1975 Body Ritual Among the Nacirema. *In* The Nacirema. James P. Spradley and Michael A. Rynkiewich, eds. pp. 10–14. Boston: Little, Brown, and Co.

Montague, Susan P., and Robert Morais

1976 Football Games and Rock Concerts: The Ritual Enactment. *In* The American Dimension: Cultural Myths and Social Realities. William Arens and Susan P. Montague, eds. pp. 33–52. Port Washington, N.Y: Alfred Publishing Co.

Myers, Alice

1977 Toward a Definition of Irony. *In* Studies in Language Variation. Ralph Fasold and Roger Shuy, eds. pp. 171–183. Washington, D.C.: Georgetown University Press.

National Bank of Detroit
1979 Employment Policy and Working Conditions. N.p.
Nye, F. Ivan, and Lois Wladis Hoffman
1963 The Employed Mother in America. Chicago: Rand McNally.
Ornstein, Robert Evans
1972 The Psychology of Consciousness. New York: Viking Press.
Pelto, Pertti J.
1970 Anthropological Research, the Structure of Inquiry. New York: Harper and Row.
Ploski, Harry, and Warren Marr
1976 The Negro Almanac: A Reference Work on the Afro-American. New York: Bellweather Co.
Porter, Sylvia
1976 Women Swell the Jobless Rolls. Gainesville Sun, August 19.
Preminger, Alex
1974 Princeton Encyclopaedia of Poetry and Poetics. Princeton: Princeton University Press.
Radcliffe-Brown, A. R.
1952 Structure and Function in Primitive Society. Glencoe, N.Y.: Free Press.
Radin, Paul
1957 Primitive Man as Philosopher. New York: Dover Publications.
Rappaport, Roy A.
1974 Obvious Aspects of Ritual. Cambridge Anthropologist 2:2–60.
Regelson, Stanley
1976 The Bagel: Symbol and Ritual at the Breakfast Table. *In* The American Dimension: Cultural Myths and Social Realities. William Arens and Susan P. Montague, eds. pp. 124–140. Port Washington, N.Y.: Alfred Publishing Co.
Reidel, Bari
1978 Rounding Out A^2. Ann Arbor: Office of Orientation, University of Michigan.
Ryan, William
1971 Blaming the Victim. New York: Vintage Books.
Schulz, Muriel
1975 The Semantic Derogation of Woman. *In* Language and Sex. Barrie Thorne and Nancy Henley, eds. pp. 64–75. Rowley, Mass.: Newbury House.
Siegal, Alberta Engvall, Lois Meek Stolz, Ethel Alice Hitchcock, and Jean Adamson
1963 Dependence and Independence in Children. *In* The Employed Mother in America. F. Ivan Nye and Lois Wladis Hoffman, eds. pp. 67–81. Chicago: Rand McNally.
Silveira, Jeanette
1972 Thoughts on the Politics of Touch. Women's Press, no. 13.
Sklar, Dusty
1976 The Trauma of Eventlessness. Family Weekly, January 11.

Smuts, Robert W.
1974 Women and Work in America. New York: Schocken Books.
Spradley, James P.
1972 Adaptive Strategies of Urban Nomads. *In* Culture and Cognition. James P. Spradley, ed. San Francisco: Chandler Publishing Co.
1979 The Ethnographic Interview. New York: Holt, Rinehart, and Winston.
Spradley, James P., and Michael A. Rynkiewich, eds.
1975 The Nacirema. Boston: Little, Brown, and Co.
State of Michigan
1978 Laws Relating to the Practice of Mortuary Science and Rules and Regulations of the Board of Examiners in Mortuary Science. Lansing, Michigan.
Stebbins, Robert
1967 A Theory of the Definition of the Situation. The Canadian Review of Sociology and Anthropology 4:148–164.
Steward, Julian
1955 Theory of Culture Change. Urbana: University of Illinois Press.
Strecker, Edward A.
1946 Their Mothers' Sons. Philadelphia: J. P. Lippincott.
Sudnow, David
1967 Passing On: The Social Organization of Dying. Englewood Cliffs, N.J.: Prentice-Hall.
Swacker, Marjorie
1975 The Sex of the Speaker as a Sociolinguistic Variable. *In* Language and Sex. Barrie Thorne and Nancy Henley, eds. pp. 76–83. Rowley, Mass.: Newbury House.
Taylor, Lawrence
1976 Coffee: The Bottomless Cup. *In* The American Dimension: Cultural Myths and Social Realities. William Arens and Susan P. Montague, eds. pp. 141–148. Port Washington, N.Y.: Alfred Publishing Co.
Thorne, Barrie, and Nancy Henley
1975 Difference and Dominance: An Overview of Language, Gender, and Society. *In* Language and Sex. Barrie Thorne and Nancy Henley, eds. pp. 5–42. Rowley, Mass.: Newbury House.
Turner, Victor W.
1974 The Ritual Process. Chicago: Aldine.
United States Department of Labor
1975 The Myth and the Reality. Washington, D.C.: U.S. Government Printing Office.
United States Savings and Loan League
1950 Suggested Outline for an Employees' Guide or Manual. N.p.
van Gennep, Arnold
1960 The Rites of Passage. Monica Vizedom and Gabrielle Caffe, transls. Chicago: University of Chicago Press (1st ed. Paris: Nourry 1909).
Whittaker, James O., and Robert Meade
1967 Sex of Communicator as a Variable in Source Credibility. Journal of Social Psychology 72:27–34.

Wolff, Kurt
1964 The Definition of the Situation. *In* The Dictionary of the Social Sciences. Julius Gould and William L. Kolb, eds. p. 182. Glencoe, N.Y.: Free Press.
Wylie, Phillip
1955 Generation of Vipers. New York: Holt, Rinehart, and Winston.
Yengoyan, Aram A.
1977 Man, Culture, and Biology. *In* Horizons of Anthropology. 2nd ed. Sol Tax and Leslie G. Freeman, eds. pp. 75–86. Chicago: Aldine.
Young, Frank W.
1962 The Function of Male Initiation Ceremonies: A Cross Cultural Test of An Alternative Hypothesis. The American Journal of Sociology 67:379–391.
Zimmerman, Don H., and Candace West
1975 Sex Roles, Interruptions and Silences in Conversation. *In* Language and Sex. Barrie Thorne and Nancy Henley, eds. pp. 105–120. Rowley, Mass.: Newbury House.
Zuckerman, Ed.
1976 Whodunit? Rolling Stone.